Lecture Notes in Mathematics

Edited by A. Dold and B. Eckmann

672

T0216085

Robert L. Taylor

Stochastic Convergence of Weighted Sums of Random Elements in Linear Spaces

Springer-Verlag
Berlin Heidelberg New York 1978

Author
Robert L. Taylor
Department of Mathematics
and Computer Science
University of South Carolina
Columbia, SC 29208/USA

Library of Congress Cataloging in Publication Data

Taylor, Robert Lee, 1943-
Stochastic convergence of weighted sums of random elements in linear spaces.

(Lecture notes in mathematics ; 672)
Bibliography: p.
Includes index.
1. Linear topological spaces. 2. Stochastic processes. 3. Limit theorems (Probability theory) 4. Law of large numbers. I. Title. II. Series: Lecture notes in mathematics (Berlin) ; 672.
QA3.L28 no. 672 [QA273.43] 510'.8s
[515'.73] 78-13024

AMS Subject Classifications (1970): 60B05, 60F15, 60G99

ISBN 3-540-08929-2 Springer-Verlag Berlin Heidelberg New York
ISBN 0-387-08929-2 Springer-Verlag New York Heidelberg Berlin

This work is subject to copyright. All rights are reserved, whether the whole or part of the material is concerned, specifically those of translation, reprinting, re-use of illustrations, broadcasting, reproduction by photocopying machine or similar means, and storage in data banks. Under § 54 of the German Copyright Law where copies are made for other than private use, a fee is payable to the publisher, the amount of the fee to be determined by agreement with the publisher.

© by Springer-Verlag Berlin Heidelberg 1978
Printed in Germany

Printing and binding: Beltz Offsetdruck, Hemsbach/Bergstr.
2141/3140-543210

PREFACE

Recent interest in representing stochastic processes as random variables in function spaces has inspired the study of the "convergence of weighted sums" for random elements. The purpose of these notes is to provide a unified presentation of the results which have been obtained to date concerning the stochastic convergence of weighted sums of random elements in (primarily) linear topological spaces, including some recent work of the author and some colleagues.

These notes are somewhat self-contained with only a background knowledge of basic probability theory being absolutely essential, Chapter I presents many of the essential definitions and results from mathematical analysis and topology that are needed, and the definitions and properties of random elements are developed in Chapter II. The remainder of the notes is concerned with the stochastic convergence results and possible applications.

Probabilists, statisticians, and mathematicians who are involved in research concerning convergence theory, stochastic processes, ergodic theory, and related topics will find these notes very useful. In addition, the notes will be useful for introducing advanced graduate students and researchers to the area of random elements and stochastic convergence of weighted sums in function spaces. It is anticipated that these notes will be expanded in a later version to include developing convergence theory results and applications.

The author is very grateful to research colleagues W. J. Padgett, Duan Wei, and Peter Daffer for work and suggestions on this manuscript and related research. A special thanks goes to Mrs. Denise Domeracki for the typing and general appearance of the manuscript. Finally, the author wishes to thank the Department of Mathematics and Computer Science at the University of South Carolina and the

Department of Statistics at the Florida State University for providing the support and resources that made these notes possible.

Robert Lee Taylor
Columbia, South Carolina
December, 1977

TABLE OF CONTENTS

GENERAL INTRODUCTION 1

I. MATHEMATICAL PRELIMINARIES 6

1.0 Introduction 6

1.1 Linear Spaces and Topologies 6

1.2 Some Basic Mathematical Tools. 12

1.3 Basis Theory 14

1.4 Problems 20

II. RANDOM ELEMENTS IN LINEAR SPACES 21

2.0 Introduction 21

2.1 Definitions and Examples of Random Elements 21

2.2 Topological Properties of Random Elements 25

2.3 Probability Properties of Random Elements 30

2.4 Problems 42

III. LAWS OF LARGE NUMBERS, UNCORRELATION, AND CONVERGENCE OF WEIGHTED SUMS OF RANDOM VARIABLES 44

3.0 Introduction 44

3.1 Laws of Large Numbers for Random Variables 45

3.2 Extensions to Separable Hilbert Spaces 49

3.3 Comparisons of Uncorrelation Concepts 54

3.4 Convergence of Weighted Sums of Random Variables . . . 63

3.5 Problems 70

IV. LAWS OF LARGE NUMBERS IN NORMED LINEAR SPACES 71

4.0 Introduction 71

4.1 Laws of Large Numbers for Independent, Identically Distributed Random Elements 72

4.2 Distributional Conditions and Laws of Large Numbers. . 81

4.3 Beck's Convexity and the Strong Law of Large Numbers . 86

4.4 Other Geometric Conditions and the Strong Law of Large Numbers 96

4.5 Other Convergence Results and Extensions to Fréchet Spaces 104

4.6 Problems 106

V. CONVERGENCE OF WEIGHTED SUMS IN NORMED LINEAR SPACES . . 108

5.0 Introduction 108

5.1 Distributional Conditions and Convergence of Weighted Sums 109

5.2 Convergence for Weighted Sums of Tight Random Elements 120

5.3 Geometric Conditions and Convergence of Weighted Sums 137

5.4 Problems 145

VI. RANDOMLY WEIGHTED SUMS 146

6.0 Introduction 146

6.1 Identically Distributed Results 146

6.2 Convergence in the rth Mean for Randomly Weighted Sums 150

6.3 Problems 152

VII. LAWS OF LARGE NUMBERS IN D[0,1] 153

7.0 Introduction 153

7.1 Preliminaries 154

7.2 Weak laws of Large Numbers in D[0,1] 166

7.3 Strong laws of Large Numbers in D[0,1] 171

7.4 A Discussion of Convex Tightness 179

7.5 Problems 184

VIII. POSSIBLE APPLICATIONS 185

8.0 Introduction 185

8.1 Applications in Stochastic Processes 186

8.2 Applications in Decision Theory 190

8.3 Applications in Quality Control 195

8.4 Applications in Estimation Problems 196

BIBLIOGRAPHY 202

SUBJECT INDEX 212

GENERAL INTRODUCTION

The consideration of a stochastic process as a random element in a function space (a random variable taking values in a function space) by Doob (1947), Mann (1951), Prohorov (1956), Billingsley (1968), and others has inspired the study of stochastic convergence properties for random elements. However, a simple example illustrates that careful construction of the appropriate framework is needed in these considerations. For example, if $\{X_t: t \in T\}$ is a stochastic process with respect to the probability space (Ω, A, P), then for each $\omega \in \Omega$ $X_t(\omega)$, $t \in T$, the sample path can be regarded as a real-valued function of $t \in T$. But in considering $\{X_t: t \in T\}$ as a mapping from Ω into R^T, the space of real-valued functions on T, certain measurability problems may occur. Define an identity function $X = \{X_t: t \in T\}$ from $\Omega = R^T$ to R^T by

$$[X_t, t \in T](\omega) = \omega \text{ for each } \omega \in R^T = \Omega.$$

Let $A = \prod_{t \in T} B(R)$ where $B(R)$ denotes the Borel subsets of R and let P be the probability measure degenerate at the origin. Then $X_{t_0}(\omega) = \omega(t_0)$ for each $t_0 \in T$ and is a random variable since

$$\{\omega: X_{t_0}(\omega) \le \alpha\} = (\prod_{t \ne t_0} R) \times (-\infty, \alpha] \in \prod_{t \in T} B(R)$$

for each $\alpha \in R$. However, $\prod_{t \in T} B(R) = B(R^T)$ if and only if T is countable [Doob (1947)]; otherwise $\prod_{t \in T} B(R) \subsetneq B(R^T)$ where the Borel field of R^T is generated by open subsets of the product topology of $R^T = \prod_{t \in T} R$. Thus, the mapping $X = \{X_t: t \in T\}$ may not be a (Borel) measureable function from (Ω, A, P) to R^T.

One way of solving the problem of measurability problem is by placing constraints on the parameter space. For example, a stochstic process with a countable parameter space will be shown to be a random element in the space of sequences, s. Often the stochastic process will take values only in a "small" subspace of R^T. Recall that a separable stochastic process may have sample paths which are Borel measurable functions from T into R [Loéve (1963), p. 510] and hence are restricted a.s. to a subspace of R^T. Thus, the stochastic processes may have properties which reduce the ranges of the mappings from Ω to interesting subspaces of R^T where different (possibly stronger) topological structures can be employed. Characterizations of stochastic properties as random elements in a variety of different spaces will be presented in Chapter II.

In an attempt to make the notes reasonably self-contained, Chapter I will consist of basic mathematical notations, definitions, and results for future reference throughout the presentation. In Chapter II the concept of a random element will be presented, and some basic properties and definitions will be given. The emphasis in Chapter II will be to illustrate the different techniques which are to be used in studying function-valued random variables. Thus, numerous examples are presented in Chapter II for the convenience of the reader and for reference in the remainder of the notes.

A weighted sum of a sequence $\{X_n\}$ of random variables is an expression of the form

$$S_n = \sum_{k=1}^{n} a_{nk} X_k \tag{0.1}$$

where $\{a_{nk}\}$ is a double array of constants. The most classical form of a weighted sum is the mean (or average) where

$$a_{nk} = \begin{cases} 1/n & \text{for } k = 1,2,\ldots,n \\ 0 & \text{for } k > n. \end{cases} \tag{0.2}$$

The convergence of $\{S_n\}$ plays a fundamental role in the relationships between the theoretical and practical aspects of both probability and statistics. With the consideration of stochastic processes as random elements in function spaces, it is very natural to seek function space analogues of the convergence results for weighted sums of random variables. Thus, Chapter III will be denoted to a brief discussion of some results on the stochastic convergence of weighted sums of random variables and some direct extensions to separable Hilbert spaces.

The law of large numbers, the convergence of $\{S_n\}$ with the particular weights listed in (0.2), was examined by Mourier (1953) where a direct extension of Kolmogorov's strong law of large numbers for independent, identically distributed random elements with finite first absolute moments was obtained for separable Banach spaces. Examples by Beck (1963) and Woyczynski (1973) showed that neither the classical second moment condition nor Kolmogorov's condition is sufficient for the strong law of large numbers in certain infinite-dimensional spaces. However, it is sufficient for certain other infinite-dimensional spaces. Such spaces have been examined by Beck (1963) (B-convex spaces) and Hoffman-Jørgensen and Pisier (1976) (Type p spaces, $1 \leq p \leq 2$). Woyczynski (1973) also showed that Chung's strong law of large numbers holds in G_α-spaces ($0 < \alpha \leq 1$). Basically, the G_α spaces provide the inequality

$$E||X_1+\ldots+X_n||^{1+\alpha} \leq C \textstyle\sum_{k=1}^{n} E||X_k||^{1+\alpha} \qquad (0.3)$$

which is valid for independent random elements $\{X_n\}$ where C is a constant. When $\alpha = 1$ in (0.3), then a bound on the variance of a sum of random elements is available. This bound plays a key role in the proofs of most results. In general, this inequality is not available. These and other laws of large numbers for random elements in normed linear spaces are presented in Chapter IV.

The classical second moment conditions alone do not yield laws of large numbers for sequences of independent random elements. An rth moment bound ($r > 1$) and the assumption of tightness do lead to laws of large number [Taylor and Wei (to appear)]. These results follow from convergence results of weighted sums without requiring any geometric structure for the spaces and are presented in Chapter V. Tightness of a sequence of random elements has been defined by LeCam (1957), and tightness characterizations of random elements in the function spaces C[0,1] and D[0,1] have been treated by Billingsley (1968). In these notes some properties of tightness are presented, and several counterexamples are provided for other plausible conjectures. Other extensions and generalizations of the results by Chow and Lai (1973), Rohatgi (1971), and Taylor (1972) are also presented in Chapter V. In particular, convergence in probability for Toeplitz weighted sums of tight random elements with an rth moment condition ($r > 1$) is shown to be equivalent for the weak linear and norm topologies of a separable Banach space. Finally, geometric conditions (type p spaces) are imposed to obtain convergence in probability results for weighted sums of random elements without assuming tightness.

Due to the random nature of many problems in the applied sciences, researchers are increasingly forced to switch from deterministic to probabilistic approaches. Thus, it is important to consider random weighting of random variables, that is, to allow $\{a_{nk}\}$ to be a double array of random variables. The results in Chapter VI pertain to these considerations. Laws of large numbers for the space D[0,1] are presented in Chapter VII. The space D[0,1] is very fruitful for applications, but with the Skorohod topology it is not a linear topological space. Hence, the results and techniques of previous chapters can not be directly applied but must be appropriately modified and developed in Chapter VII.

In the last chapter, some examples are given to demonstrate possible applications of these convergence results. Since the area of stochastic processes provided the motivation for the study of random elements, many of the applications will be concerned with stochastic processes. In addition, weighted sums often occur in the fields of applied statistics such as robust analysis, regression analysis, control charts, decision theory, and estimation. Specific examples are given in Chapter VIII to indicate possible applications in these areas.

CHAPTER I

MATHEMATICAL PRELIMINARIES

1.0 INTRODUCTION

For the convenience of the reader and for future reference, this chapter will contain some of the basic mathematical definitions, theorems, and notations which will be used throughout the notes. Section 1.1 will present the concept of a linear topological space and will give the notation for particular linear topological spaces which will be used in later chapters. In Section 1.2 some basic theorems and results are listed for later reference. The definition of a Schauder basis and some of its properties will be given in Section 1.3. Basis theory provides very useful tools for characterizing random elements in linear topological spaces and for obtaining convergence results. A few problems are listed at the end of the chapter and can be used to check on understanding the basic concepts of this chapter.

1.1 LINEAR SPACES AND TOPOLOGIES

The following definitions and notation will be used throughout these notes. First, R will denote the real numbers.

Definition 1.1.1 A nonempty set X is said to be a (real) linear space if there is defined an operation of addition which makes X a commutative group and an operation of multiplication by scalars (real numbers) which satisfies the distributive, and identity laws; more precisely stated,

(1) to every pair of elements $(x,y) \in X \times X$ there corresponds an element $z \in X$ such that $z = x + y$,

(2) to every $x \in X$ and $t \in R$ there corresponds an element $tx \in X$,

(3) the operations defined in (1) & (2) satisfy the following properties for every $x,y,z \in X$ and $s,t \in R$

(i) $x + y = y + x$

(ii) $(x+y) + z = x + (y+z)$

(iii) $x + y = x + z$ implies $y = z$

(iv) $1x = x$

(v) $s(tx) = (st)x$

(vi) $(s+t)x = sx + tx$

(vii) $s(x+y) = sx + sy$.

Note that Definition 1.1.1 implies the existence of a zero element, 0, and an additive inverse, $-x$.

Definition 1.1.2 A nonempty set M is called a semimetric space if there is a real-valued function d defined on $M \times M$ with the following properties:

(i) $d(x,y) = d(y,x) \geq 0$ for all $(x,y) \in M \times M$;

(ii) $d(x,x) = 0$ for all $x \in M$; and

(iii) $d(x,z) \leq d(x,y) + d(y,z)$ for all $x,y,z \in M$.

If in addition

(iv) $d(x,y) = 0$ if and only if $x = y$, then M is called a metric space. The real number $d(x,y)$ is called the distance from x to y, and d is called a semimetric (or a metric when (iv) holds).

Definition 1.1.3 A sequence $\{x_n\}$ in a metric space M is called a Cauchy sequence if for every $\varepsilon > 0$ there exists an integer N such that $d(x_n, x_m) < \varepsilon$ whenever $n \geq N$ and $m \geq N$.

Definition 1.1.4 A metric space M is said to be complete if every Cauchy sequence in M converges to an element of M.

Definition 1.1.5 If X is a linear space with a topology such that the two basic operations of addition and scalar multiplication are continuous, then X is said to be a linear topological space or a topological vector space.

It is not sufficient for a linear space to have a topology defined on it in order to be a linear topological space. For example, consider the real numbers R and the discrete metric topology.

Definition 1.1.6 The collection of all continuous linear functionals (that is, continuous linear real-valued functions) defined on a linear topological space X is called the dual space of X and will be denoted by X*.

If a linear topological space X has a metric d which generates its topology, then X is called a linear metric space. Also, the concept of a seminorm or norm is often used to introduce a topology on a linear space. A detailed discussion of seminorms and norms may be found in Yosida (1965) and Wilansky (1964).

Definition 1.1.7 A real-valued function p defined on a linear space X is said to be a seminorm on X if the following properties are satisfied for all $x,y \in X$ and $t \in R$;

(i) $p(x) \geq 0$;

(ii) $p(x+y) \leq p(x) + p(y)$; (subadditivity)

(iii) $p(tx) = |t|p(x)$ (homogeneity).

Definition 1.1.8 A linear space X is said to be normed if there is a real-valued function defined on X and denoted by $|| \ ||$ such that $|| \ ||$ satisfies (i), (ii), and (iii) of Definition 1.1.7 and $||x|| = 0$ if and only if $x = 0$. The function $|| \ ||$ is then called a norm.

A norm (or seminorm) always yields a topology on a linear

space where both addition and scalar multiplication are continuous operations. A complete normed linear space is called a Banach space. A complete linear metric space is called a Fréchet space. Note that a Fréchet space (by this definition) need not be locally convex while a normed linear space is locally convex.

Definition 1.1.9 Let X be a linear space. A mapping $\langle \cdot , \cdot \rangle$ from $X \times X$ into the real (or complex) numbers is called an inner product if for each $x,y,z \in X$ and t R the following properties hold:

(i) $\langle x+y,z\rangle = \langle x,z\rangle + \langle y,z\rangle$;

(ii) $\langle tx,y\rangle = t\langle x,y\rangle$;

(iii) $\overline{\langle x,y\rangle} = \langle y,x\rangle$, the bar denotes the complex conjugation;

(iv) $\langle x,x\rangle > 0$ if and only if $x \neq 0$.

The space X is then said to be an inner product space. If X with the norm defined by $||x|| = \langle x,x\rangle^{1/2}$ is complete, then X is called a Hilbert space.

Definition 1.1.10 A subset S of a topological space X is said to be dense in X if its closure (the smallest closed set containing S) equals X. If X has a countable dense subset, then X is called a separable space.

Most of the spaces that are discussed in these notes are separable since most results for random elements require separable spaces. Several particular linear topological will now be listed for notational reference [see Taylor (1958), Wilansky (1964), or Yosida (1965) for further discussions of these spaces].

(1) The space of all real sequences will be denoted by s. A metric on s can be defined by the Fréchet metric

$$d(x,y) = \sum_{k=1}^{\infty} 2^{-k} |x_k - y_k| / (1 + |x_k - y_k|)$$

for every $x = (x_1,x_2,\ldots)$ and $y = (y_1,y_2,\ldots)$ in s. With this metric s is a separable Fréchet space. Moreover, for each k, $p_k(x) = |x_k|$ defines a seminorm.

(2) The space of all real convergent sequences

$$\{x = (x_1,x_2,\ldots)\colon \lim_{n\to\infty} x_n \text{ exists}\}$$

will be denoted by c. A norm for c is defined by $||x|| = \sup_n |x_n|$. With this norm c is a separable Banach space. Similarly, $c_0 \subset c$ will denote the separable Banach space of all null convergent sequences with norm $||x|| = \sup_n |x_n|$.

(3) The symbol $R^{(\infty)}$ will denote the subspace of c_0 of all real sequences which are zero except for a inite number of coordinates. With norm defined by $||x|| = \sup_n |x_n|$, the space $R^{(\infty)}$ is an incomplete normed linear space.

(4) The spaces ℓ^p, $1 \le p < \infty$, are the spaces of all sequences $x = (x_1,x_2,\ldots)$ such that $\sum_{k=1}^{\infty} |x_k|^p < \infty$. With the norm defined by

$$||x|| = (\sum_{k=1}^{\infty} |x_k|^p)^{1/p},$$

each of the spaces ℓ^p, $1 \le p < \infty$, is a separable Banach space. For $p = 2$, ℓ^2 is a Hilbert space with inner product

$$<x,y> = \sum_{k=1}^{\infty} x_k y_k .$$

(5) The symbol ℓ^∞ will denote the space of all bounded sequences with the norm defined by $||x|| = \sup_n |x_n|$. The space is a Banach space but is not separable.

(6) The space of all real-valued continuous functions on the interval $[0,1]$ will be denoted by $C[0,1]$ with the norm defined by

$$||x|| = \sup_{0\le t\le 1} |x(t)| \quad \text{for } x \in C[0,1].$$

The space $C[0,1]$ is also a separable Banach space.

(7) The spaces of all real Lebesgue measurable functions x on an interval [a,b] such that

$$\int_a^b |x(t)|^p dt < \infty$$

will be denoted by $L^p[a,b]$, $1 \le p < \infty$. The norm of $L^p[a,b]$ is defined by

$$||x|| = (\int_a^b |x(t)|^p dt)^{1/p} \quad \text{(almost everywhere identifications).}$$

The $L^p[a,b]$ spaces are separable Banach spaces, and $L^2[a,b]$ is a Hilbert space. When the interval is all of the real line, then the spaces are simply denoted by L^p.

(8) The symbol $L^\infty[a,b]$ will denote the space of all functions that are Lebesgue measurable and bounded almost everywhere on [a,b] with the norm defined by

$$||x|| = \beta(|x|) = \inf \{\delta: \ |x(t)| \le \delta \ \text{ a.e.}\}$$

where $\beta(|x|)$ is called the essential supremum of $|x(t)|$. Similar to ℓ^∞, $L^\infty[a,b]$ is a Banach space which is not separable when $a < b$.

(9) Let $C[0,\infty)$ denote the space of all continuous real-valued functions on $[0,\infty)$. Define a sequence of seminorms on $C[0,\infty)$ by

$$p_k(x) = \sup_{0 \le t \le k} |x(t)| \quad \text{for } k = 1,2,\ldots .$$

Then a metric d can be defined on $C[0,\infty)$ by the Fréchet combination of the seminorms

$$d(x,y) = \sum_{k=1}^\infty 2^{-k} p_k(x-y)/[1+p_k(x-y)]. \tag{1.1.1}$$

With this metric $C[0,\infty)$ is a separable Fréchet space [Whitt (1970)].

(10) The symbol F will denote a linear space with a countable family of seminorms $\{p_k\}$ such that for each $x \in F$ $p_k(x) = 0$ for all k implies that $x = 0$. If a metric is defined by Equation (1.1.1), then F is a locally convex Fréchet space [see Yosida (1965),

pp. 24-26]. Note that the spaces s and $C[0,\infty)$ defined in (1) and (9) are such Fréchet spaces. For each k the linear space F with the seminorm p_k is a seminormed linear space which will be denoted by F_k. The metric topology on F is stronger than the seminorm topology. However, convergence in the metric topology of F is equivalent to convergence in every seminormed space F_k, $k = 1,2,\ldots$.

1.2 SOME BASIC MATHEMATICAL TOOLS

In this section some miscellaneous definitions, notations, and results will be stated. These will be frequently used in proving convergence results for random elements.

Let X be a topological space and let B(X) denote the smallest sigma-field which contains the open subsets of X, that is, B(X) denotes the Borel subsets of X. A function T from X into a topological space Y is said to be Borel measurable if for each Borel set $A \in B(Y)$,

$$T^{-1}(A) = \{x \in X: \ T(x) \in A\} \in B(X).$$

Proposition 1.2.1 Let X be a separable semimetric space with semimetric d. For each $\lambda > 0$ there exists a countably-valued Borel measurable function $T : X \to X$ such that $d(T_\lambda(x),x) < \lambda$ for each $x \in X$.

Proof: Since X is separable, choose a countable dense subset $\{x_1,x_2,\ldots\}$. For $\lambda > 0$ form the countable collection of λ-neighborhoods $N(x_i) = \{x: \ d(x,x_i) < \lambda\}$ which covers X. Define the countably valued Borel measurable function T_λ by

$$T_\lambda(x) = x_1 \quad \text{if } x \in N_\lambda(x_1)$$

and

$$T_\lambda(x) = x_n \text{ if } x \in N_\lambda(x_n) - [\bigcup_{i=1}^{n-1} N(x_i)]$$

for n = 2,3,... . ///

The laws of large numbers for linear topological spaces were first proved for countably-valued random elements and then extended to arbitrary random elements by using Proposition 1.2.1. Other results have been obtained by using embedding techniques (see Padgett and Taylor (1973)). Basic to these techniques is the concept of an isometry.

Definition 1.2.1 Let X and Y be linear topological spaces. A one-to-one bicontinuous linear function from X onto Y is called an isomorphism. The two spaces X and Y are said to be isomorphic if there is an isomorphism from each space onto the other.

Let M and M_1 be linear semimetric spaces with semimetrices d and d_1 respectively. A function T: $M \to M_1$ is called an isometry if $d_1(T(x),T(y)) = d(x,y)$ for all $x,y \in M$. The two spaces M and M_1 are said to be isometric if there is an isometry from each space onto the other.

The composition of two functions ϕ and γ will be denoted by $\phi \circ \gamma$. That is, if γ: $X \to Y$ and ϕ: $Y \to Z$ where X, Y, and Z are linear spaces, then for each $x \in X$, $(\phi\circ\gamma)(x) = \phi(\gamma(x)) \in Z$.

Let X be a seminormed linear space with seminorm p and let $f \in X^*$. Define $||f|| = \sup\{|f(x)|: p(x) \le 1\}$. Finally, the Hahn-Banach Theorem and a useful corollary will be stated since it is of fundamental importance in developing properties of random elements.

Theorem 1.2.2 (Hahn-Banach) Let S be a subspace of a linear space X and let p be a seminorm defined on X. Suppose that f is a linear functional defined on S such that $|f(x)| \le p(x)$ for all $x \in S$.

Then there is a linear functional f_1 on X such that $f_1(x) = f(x)$ for all $x \in S$ and $|f_1(x)| \leq p(x)$ for all $x \in X$.

Corollary 1.2.3 [Wilansky (1964), p. 67] Let X be a seminormed linear space and $x \in X$ such that $p(x) \neq 0$. Then there exists an $f \in X^*$ such that $||f|| = 1$ and $f(x) = p(x)$.

1.3 BASIS THEORY

The definition and some basic properties of a Schauder basis (or simply a basis) for a linear topological space are presented in this section. Several convergence results in Chapters IV and V are proved for normed linear spaces or Banach spaces which have Schauder bases. Particular bases will be given for some of the spaces which were defined in Section 1.1.

Definition 1.3.1 Let X be a linear topological space. A sequence $\{b_n\} \subset X$ is a Schauder basis for X if for each $x \in X$ there exists a unique sequence of scalars $\{t_n\}$ such that

$$x = \lim_{n\to\infty} \sum_{k=1}^{n} t_k b_k.$$

A Schauder basis $\{b_n\}$ for a linear metric space is called monotone if the sequence of real numbers

$$\{d(\sum_{k=1}^{n} t_k b_k, 0): \quad n = 1,2,3,\ldots\}$$

is monotone increasing for each sequence of scalars $\{t_n\}$.

When a linear topological space X has a basis $\{b_n\}$, a sequence of linear functionals $\{f_k\}$ can be defined by letting $f_k(x) = t_k$, $k = 1,2,\ldots$, where $x \in X$ and

$$x = \lim_{n\to\infty} \sum_{k=1}^{n} t_k b_k.$$

The linear functionals $\{f_k\}$ are called the coordinate functionals (for the basis $\{b_n\}$). The coordinate functionals depend on the basis and need not be continuous (see Example 1.3.1). However, as a consequence of the open-mapping theorem (or Lemma 1.3 1), the coordinate functionals of a Banach space are continuous. Finally, a sequence of linear functions $\{U_n\}$ on X can be defined by letting

$$U_n(x) = \sum_{k=1}^{n} f_k(x) b_k$$

for each $x \in X$. The sequence of functions $\{U_n\}$ is called the sequence of partial sum operators (for the basis $\{b_n\}$).

Lemma 1.3.1 (a) If X is a normed linear space which has a monotone basis, then $||U_n|| \leq 1$ for each n, that is, $||U_n(x)|| \leq ||x||$ for each $x \in X$ and each n.

(b) If X is a Banach space which has a Schauder basis, then there exists a positive constant m such that $||U_n|| \leq m$ for m. (The constant m is referred to as the basis constant.)

Proof - Part (b) [Condensed version of proofs which are given in Chapter 11 of Wilansky (1964)]: Let $\{b_n\}$ be a Schauder basis for a Banach space X. Define a norm p on X by letting $p(x) = \sup_n\{||U_n(x)||\}$ for each $x \in X$. Thus, $\{b_n\}$ is a monotone basis with respect to the norm p, and X is a Banach space with the topology of the norm p. The two norms for X are equivalent by the open-mapping theorem. Thus, there exists a constant m such that $p(x) \leq m||x||$ for all $x \in X$, and hence $||U_n(x)|| \leq \sup_n ||U_n(x)|| = p(x) \leq m||x||$. ///

Example 1.3.1: Let X be the space of real polynomials with domain [0,1] and norm $||x|| = \sup_{0 \leq t \leq 1} |x(t)|$. The set $\{b_n = t^n : n = 0,1,2,\ldots\}$ forms a Schauder basis, and the coordinate functionals are given by

$$f_k(x) = \frac{d^k x(t)}{dt^k}\bigg|_{t=0} \frac{1}{k!} .$$

Let $y_n(t) = (1-t)^n$. Note that $||y_n|| \le 1$ for each n and that $|f_1(y_n)| = n$. Hence, f_1 is not continuous since $||\frac{1}{n}y_n|| \to 0$ as $n \to \infty$ but $|f_1(\frac{1}{n}y_n)| = 1$ for all n. Only the 0th coordinate functional $f_0(x) = x(0)$ is continuous. Lemma 1.3.1 does not apply since X is not complete. ///

The space c_0, ℓ^p, and s have $\{\delta^n\}$ as a Schauder basis where $\delta^1 = (1,0,0,\ldots)$, $\delta^2 = (0,1,0,\ldots)$, ... The space c has $\{(1,1,1,\ldots), \delta^1, \delta^2, \ldots\}$ as a Schauder basis. A basis for C[0,1] is given in Example 1.3.2. See Wilansky (1964) and Marti (1969) for further discussions of Schauder bases and their properties.

Example 1.3.2: For the space C[0,1], a Schauder basis and a set of coordinate functionals are defined as follows [Marti (1969)]. Let $b_n(t) = 0$ if $t \notin [0,1]$. Let

$$b_0(t) = t \qquad\qquad f_0(x) = x(1)$$

$$b_1(t) = 1-t \qquad\qquad f_1(x) = x(0)$$

$$b_2(t) = \begin{cases} 2t & \text{if } 0 \le t \le \frac{1}{2} \\ 2-2t & \text{if } \frac{1}{2} \le t \le 1 \end{cases} \qquad\qquad f_2(x) = x(\tfrac{1}{2}) - \frac{x(0) + x(1)}{2}$$

$$b_3(t) = b_2(2t) \qquad\qquad f_3(x) = x(\tfrac{1}{4}) - \frac{x(\frac{1}{2}) + x(0)}{2}$$

$$b_4(t) = b_2(2t-1) \qquad\qquad f_4(x) = x(\tfrac{3}{4}) - \frac{x(\frac{1}{2}) + x(1)}{2} .$$

In general,

$$b_{2^n+i}(t) = b_2(2^n t - i + 1), \quad i = 1,\ldots,2^n$$

and

$$f_{2^n+i}(x) = x(\tfrac{2i-1}{2^{n+1}}) - \frac{x(\frac{i-1}{2^n}) + x(\frac{i}{2^n})}{2}, \quad i = 1,\ldots,n. \qquad ///$$

In examining applications for continuous sample path stochastic processes, it is necessary to list an extension of the traditional Schauder basis for C[0,1] (as given in Example 1.3.2) to a Schauder basis for $C[0,\infty)$. First, define

$$b_{1,0}(t) = \begin{cases} 1-t & \text{if } 0 \le t \le 1 \\ 0 & \text{if } t \in (1,\infty) \end{cases} \qquad f_{1,0}(x) = x(0),$$

$$b_{1,1}(t) = \begin{cases} t & \text{if } 0 \le t \le 1 \\ 2-t & \text{if } 1 < t \le 2 \\ 0 & \text{if } t \in (2,\infty) \end{cases} \qquad f_{1,1}(x) = x(1),$$

and in general

$$b_{1,k}(t) = \begin{cases} b_{1,1}(t-k+1) & \text{if } k-1 \le t \le k+1 \\ 0 & \text{if } t \notin [k-1,k+1] \end{cases} \quad f_{1,k}(x) = x(k) \quad .$$

Next, for each k and each $m \ge 2$ define $b_{m,k}(t) = b_{k+1}(t-m+2)$ where $\{b_k(u)\}$ is the traditional Schauder basis for $C[0,1]$ in Example 1.3.2. Thus, for $m \ge 2$

$$\{b_{1,m-2},\ b_{1,m-1},\ b_{m,1},\ b_{m,2},\ldots\}$$

is a translate of the traditional Schauder basis when restricted to the interval $[m-2,m-1]$. Moreover, the coordinate functional corresponding to the element $b_{m,(2^i+j-1)}(t)$ is

$$f_{m,(2^i+j-1)}(x) = x(m-2+\frac{2j-1}{2^{i+1}}) - \frac{x(m-2+\frac{j-1}{2^i})+x(m-2+\frac{j}{2^i})}{2}.$$

Finally, relabel the elements and corresponding functionals $c_0,c_1,c_2,\ldots,$ and $g_0,g_1,g_2,\ldots,$ so that for each positive integer $\frac{n(n+1)}{2}$, n+1 of the elements $\{c_0,c_1,c_2,\ldots,c_{\frac{n(n+1)}{2}}\}$ are nonzero on $[0,1]$ and n-k+1 of the elements are nonzero on $[k,k+1]$ for each $k = 1,2,3,\ldots,n$.

From the preceding construction, for each $x \in C[0,\infty)$

$$\lim_{n\to\infty} \left(\sup_{0\le t\le \ell} \left| x(t) - \textstyle\sum_{i=0}^n g_i(x)c_i(t) \right| \right) \to 0$$

for each positive integer ℓ. Moreover,

$$p_\ell(\sum_{i=0}^{n} g_i(x)c_i)$$

is an increasing sequence in n for each ℓ. Hence, $\{c_0, c_1, c_2, \ldots\}$ is a monotone Schauder basis for $C[0,\infty)$, and $\{g_0, g_1, g_2, \ldots\}$ are the coordinate functionals.

Two elements x and y in a Hilbert space H are said to be orthogonal or perpendicular to each other if $\langle x,y\rangle = 0$. A set of elements of H is called an orthogonal set if every pair of elements in the set is orthogonal. An orthogonal set of unit (norm) vectors $\{b_\alpha\}$ in H is called an orthonormal set. It will be said that an orthonormal set S in H is maximal whenever there is no point of H (except the zero element) which is orthogonal to all elements of S. The following results concerning orthonormal sets in a Hilbert space will be needed in Chapters II and III.

Lemma 1.3.2 Any maximal orthonormal subset S of a Hilbert space H is an orthonormal basis for H in the sense that every element of H is a unique (possibly infinite) linear combination of elements of S.

The proof of Lemma 1.3.2 follows from the basic result that if S is an orthonormal set in a Hilbert space H, then for any $a \in H$ the series

$$\sum_{b \in S} \langle a,b\rangle b$$

is convergent to a_0 and $a-a_0$ is orthogonal to S. Thus, if S is a maximal orthonormal subset of H, then for any $a \in H$,

$$a = \sum_{b \in S} \langle a,b\rangle b.$$

Remark 1.3.1 Therefore, any countable maximal orthonormal set, that is, a countable orthonormal basis $\{b_n\}$, is a Schauder basis for the Hilbert space H. Furthermore, for $x,y \in H$

$$\langle x,y\rangle = \sum_{n=1}^{\infty} \langle x,b_n\rangle\overline{\langle y,b_n\rangle}$$

from Wilansky (1964), p. 132.

The last result of this section characterizes compact sets in a Banach space which has a Schauder basis as being almost finite-dimensional. This result is especially instructive since many of the convergence results are for tight random elements and tightness is defined in terms of compact sets.

Lemma 1.3.3 [Taylor and Wei (to appear)] Let K be a compact set in a Banach space which has a Schauder basis. For each $\eta > 0$ there exists a positive integer N such that $||x-U_n(x)|| < \eta$ for all $x \in K$ and $n \geq N$.

Proof: Let $g_n(x) = \sup_{k\geq n}||x-U_k(x)||$. Then

$$|g_n(x)-g_n(y)| \leq \sup_{k\epsilon n}|\ ||x-U_k(x)|| - ||y-U_k(y)||\ |$$

$$\leq \sup_{k\geq n}||(x-y) - U_k(x-y)|| \leq (m+1)||x-y||$$

where m is the basis constant. Hence, g_n is uniformly continuous for each n. Moreover, $\{g_n\}$ decreases monotonically (pointwise) to zero, and hence converges uniformly to zero by Dini's Theorem (Royden (1972), p. 162). ///

Since $U_n(K)$ remains a compact set, Lemma 1.3.2 states that K is within an "η" of being a closed and bounded subset of a finite-dimensional subspace. Moreover, the converse is sufficient for relative compactness when $\sup_{x\epsilon K}||x|| < \infty$. To obtain the converse, it suffices to show that the set is totally bounded since the space is complete. To form a finite ε-net, let t be such that $||x-U_t(x)|| < \varepsilon/2$ for all $x \in K$. Next, $\sup_{x\epsilon K}||U_t(x)|| \leq m \sup_{x\epsilon K}||x|| < \infty$ implies that there exists $x_1,\ldots,x_n \in U_t(X)$ such that

$$\{U_t(x):\ x \in K\} \subset \bigcup_{i=1}^{n} N(x_i, \varepsilon/2).$$

Hence, $||x-x_i|| \leq ||x-U_t(x)|| + ||U_t(x)-x_i||$ implies that

$$K \subset \bigcup_{i=1}^{n} N(x_i, \varepsilon).$$

1.4 PROBLEMS

1.1 Using Definition 1.1.1, verify the existence of the zero element and the additive inverse.

1.2 Show that the real numbers with the discrete metric topology is not a linear topological space.

1.3 Show that a norm on a linear space defines a topology which makes the space a linear topological space.

1.4 Show that $R^{(\infty)}$ is not complete by defining a Cauchy sequence which does not converge to an element in $R^{(\infty)}$.

1.5 Show that convergence in the metric topology of F is equivalent to convergence in all the seminormed spaces, F_k's.

1.6 Let X be a Banach space. Using Lemma 1.3.1, show that each coordinate functional of a Schauder basis for X is continuous. (Hint: need only to show $|f_n(x)| \leq C||x||$ for some $C \in R$).

CHAPTER II

RANDOM ELEMENTS IN LINEAR SPACES

2.0 INTRODUCTION

This chapter will be concerned with the definition and examples of random elements and their properties which are necessary for the study of stochastic convergence of weighted sums. When possible the definitions and results will be given for topological spaces and linear spaces which, of course, include linear metric spaces and Banach spaces. If a particular definition or result requires certain types of linear topological spaces such as separable Banach spaces, it will be so stated.

Throughout this chapter X will denote a topological space and d will denote a semimetric. The class of Borel subsets of X will be denoted by B(X); that is, B(X) will denote the smallest sigma-field containing the open subsets of X.

2.1 DEFINITIONS AND EXAMPLES OF RANDOM ELEMENTS

Let (Ω,A,P) be a probability space. A random element in X will be defined as a measurable function from (Ω,A,P) into X.

<u>Definition 2.1.1</u> A function $V: \Omega \to X$ is said to be a random element in X if $\{\omega \in \Omega: V(\omega) \in B\} \in A$ for each $B \in B(X)$.

A random element is a generalization of a random variable since the sigma-field generated by all intervals of real numbers of the form $(-\infty,b]$ is the class of Borel subsets of the real numbers R. Thus, V is a random element in R if and only if V is a random variable. Furthermore, random elements in n-dimensional Euclidean space R^n are random vectors or n-dimensional random variables. A few basic properties and lemmas are needed before considering examples of random elements in the linear topological spaces which were listed in Chapter I. These properties are generalizations of properties for random variables.

Lemma 2.1.1 If V is a random element in X and T is a Borel measurable function from X into a topological space Y, then $T(V) \equiv T \circ V$ is a random element in Y.

Proof: Let $B \in B(Y)$. Since T is a Borel measurable function, $T^{-1}(B) \in B(X)$. Thus,

$$(T \circ V)^{-1}(B) = V^{-1}(T^{-1}(B)) \in A$$

since V is a random element. ///

Property 2.1.2 Let $\{E_n\} \subset A$ be a sequence (possibly finite) of disjoint sets such that $\bigcup_n E_n = \Omega$. If $\{x_n\}$ is a sequence of elements in X and V is a function from Ω into X such that $V(\omega) = x_n$ whenever $\omega \in E_n$, then V is a random element in X.

Proof: If $A \in B(X)$ and $\{x_{n_j}\}$ is the set of elements of $\{x_n\}$ which are in B, then $V^{-1}(B) = \bigcup_j E_{n_j} \in A$. ///

Property 2.1.3 Let $\{V_n\}$ be a sequence of random elements in a semimetric space M such that $V_n(\omega) \to V(\omega)$ for each $\omega \in \Omega$. Then V is a random element in M.

Proof: It is sufficient to show that $V^{-1}(C) \in A$ for every closed subset C of M. For each positive integer k let

$$C_k = \bigcup_{x \in C} N(x,1/k),$$

where $N(x,1/k) = \{y \in M: d(x,y) < 1/k\}$. Then C_k is an open set, and

$$V^{-1}(C) = \bigcap_{k=1}^{\infty} \bigcup_{n=1}^{\infty} \bigcap_{m=n}^{\infty} V_m^{-1}(C_k). \tag{2.1.1}$$

Since V_m is a random element in M for each m and C_k is an open set, $V_m^{-1}(C_k) \in A$. Hence, $V^{-1}(C) \in A$ by Equation (2.1.1). (Note that M need not be separable.) ///

Properties 2.1.2 and 2.1.3 can be used to show that every random element in a separable semimetric space M is the uniform limit of a sequence of countably-valued random elements.

Property 2.1.4 Let M be a separable semimetric space. A mapping $V: \Omega \to M$ is a random element if and only if there exists a sequence $\{V_n\}$ of countably-valued random elements which converge uniformly to V.

Proof: By Property 2.1.3 V is a random element if there exists a sequence of countably-valued random elements converging uniformly to V.

To show the 'only if' part let $V_n = T_n \circ V$ where T_n is the countably-valued Borel measurable function T_n on M such that $d(T_n(x),x) < 1/n$ for all $x \in M$ (see Proposition 1.2.1). By Lemma 2.1.1 V_n is a random element in M. Moreover, by construction it is countably-valued, and $d(V_n,V) < 1/n$ uniformly. Thus, V_n converges to V uniformly. ///

Many writers define a random element in a Banach space as a strongly measurable function from a probability space (Ω,A,P) to the Banach space. A function $V: \Omega \to X$ is said to be strongly measurable if there exists a sequence of countably-valued measurable functions V_n such that $\lim_{n\to\infty} ||V_n(\omega)-V(\omega)|| = 0$ pointwise (or almost surely). For separable Banach spaces, Property 2.1.4 shows that the two definitions of a random element are (almost surely) the same. For nonseparable Banach spaces the range of a strongly measurable function must be (almost surely) a separable subset.

Lemma 2.1.5 If V is a random element in a topological space X and A is a random variable, then AV is a random element in X.

Proof: Let $\{A_n\}$ be a sequence of countably-valued random variables converging pointwise to A. By the continuity of scalar multiplication, A_nV is a random element in X for each n. Since A_nV converges pointwise to AV, AV is a random element in X by Property 2.1.3. ///

Let X be a linear metric space which has a Schauder basis $\{b_n\}$ such that the coordinate functionals $\{f_n\}$ are continuous. Random elements in X can easily be characterized in terms of the basis and the coordinate functionals. Let V be a random element in X. Then

$$V = \sum_{k=1}^{\infty} f_k(V) b_k \tag{2.1.2}$$

pointwise. Thus, for the sequence spaces s, c_0, and ℓ^p $(p \geq 1)$, each random element V is expressable as a sequence of random variables $\{f_k(V)\}$ (Lemma 2.1.1 implies $f_k(V)$ is a random variable since f_k is continuous and hence Borel measurable), that is,

$$V = (f_1(V),\ f_2(V), \ldots) = (V_1, V_2, \ldots)$$

where a basis $\{b_n\}$ is $\delta^1 = (1,0,0,\ldots)$, $\delta^2 = (0,1,0,\ldots), \ldots$.
A random element may be constructed by the use of the Schauder basis Let $\{A_k\}$ be a sequence of random variables such that

$$\lim_{n\to\infty} \sum_{k=1}^{n} A_k(\omega) b_k \tag{2.1.3}$$

exists in X for each $\omega \in \Omega$, for example, if X is an ℓ^p space the random variables $\{A_n\}$ must be p-summable. Lemma 2.1.5 provides that each term $A_1b_1, A_2b_2, \ldots, A_nb_n$ is a random element. Lemma 2.2.1 (of the next section) will show that the sum is a random element.

Hence, Property 2.1.3 provides that the limit in (2.1.3) is a random element in X. Thus, the random elements in X are completely characterized.

For the space C[0,1], V: $\Omega \rightarrow C[0,1]$ is a random element if and only if V_t is a random variable for each $t \in [0,1]$ (Billingsley (1968), page 57). Whitt (1970) has extended this result to $C[0,\infty)$.

2.2 TOPOLOGICAL PROPERTIES OF RANDOM ELEMENTS

Not all of the properties of random variables can be extended to random elements in topological spaces. For example, sums of random variables are random variables, but sums of random elements in X may not be defined. Even when considering linear semimetric spaces, separability is often needed to extend the basic properties of random variables to random elements. In this section the topological properties of random elements are examined to determine how they help or hinder the development of the theory of random elements.

Many of the results concerning random elements in a linear semi metric space M depend on the fact that d(V,Z) is a random variable whenever V and Z are random elements in M. In many spaces the semimetric d is induced by a norm or is the Fréchet combination of a sequence of seminorms.

Lemma 2.2.1 For a separable semimetric space M with semimetric d, d(V,Z) is a random variable whenever V and Z are random elements in M.

Proof: Let $B(M) \times B(M)$ be the sigma-field generated by sets of the form $E \times F$ where $E,F \in B(M)$. Billingsley (1968), p. 225, has shown that $B(M \times M) = B(M) \times B(M)$ when M is separable. Hence, the function (V,Z): $\Omega \rightarrow M \times M$ defined by $(V,Z)(\omega) = (V(\omega),Z(\omega))$

is a random element in $M \times M$. Since $d: M \times M \to R$ is a continuous mapping, $d(V,Z)$ is a random variable by Lemma 2.1.1. ///

If M is not separable, then $d(V,Z)$ may not be a random variable. For example, let X be the linear space of all bounded real-valued functions on R. Let Y be the space of all functions from R into $\{0,1\}$. Then, card $(X) \geq$ card $(Y) = 2^c > c$ where c is the cardinality of the continuum. Let a norm on X be defined by $||x|| = \sup_t |x(t)|$.

For $A \in B(X) \times B(X)$, let

$$P(A) = \begin{cases} 1 & \text{if } (0,0) \in A \\ 0 & \text{if } (0,0) \notin A \end{cases}$$

Then $(X \times X, B(X) \times B(X), P)$ is a probability space. Now define random elements V and Z from $X \times X$ into X by

$$V(f,g) = f \quad \text{and} \quad Z(f,g) = g \tag{2.2.1}$$

for $(f,g) \in X \times X$. Then $d(V,Z) = ||V-Z||$ is not a random variable since $\{0\} \in B(R)$ and

$$\begin{aligned} d(V,Z)^{-1}(\{0\}) &= \{(f,g)\colon \ d(V(f,g),Z(f,g)) = 0\} \\ &= \{(f,g)\colon \ |f(t) - g(t)| = 0 \text{ for all } t \in R\} \\ &= \{(f,f)\colon \ f \in X\} \notin B(X) \times B(X) \end{aligned}$$

since the cardinality of X exceeds that of the continuum and $B(X)$ contains the one point sets [Nedoma (1957)].

If X is a seminormed linear space with a seminorm p, then by Lemma 2.1.1 $p(V)$ is a random variable whenever V is a random element in X. If V is a random element in a linear topological space X, then $f(V)$ is a random variable for each $f \in X^*$. Moreover, when X is a separable seminormed linear space, the following lemma shows that the converse of the above statement is true.

Lemma 2.2.2 If X is a separable seminormed linear space, then a function $V\colon\ \Omega \to X$ is a random element if and only if $f(V)$ is a random variable for each $f \in X^*$.

Proof ("if" part): Assume $f(V)$ is a random variable for each $f \in X^*$. Let $B(C) = \sigma\{f^{-1}(B)\colon\ f \in X^* \text{ and } B \in B(R)\}$. It suffices to show that $B(C) = B(X)$ since $V^{-1}(f^{-1}(B)) = (f(V))^{-1}(B) \in A$ for each $B \in B(R)$. Note that clearly $B(C) \subset B(X)$.

To establish $B(X) \subset B(C)$, let $\{x_n\}$ be a countable dense subset of X. By Corollary 1.2.3 to the Hahn-Banach Theorem choose a sequence $\{f_n\} \subset X^*$ such that $||f_n|| = 1$ and $f_n(x_n) = p(x_n)$ for $n = 1,2,\ldots$. For $r > 0$ let $C_1 = \{x\colon\ p(x) \le r\}$ and

$$C_2 = \bigcap_{n=1}^{\infty} \{x\colon\ |f_n(x)| \le r\}.$$

If $p(x) \le r$, then $|f_n(x)| \le ||f_n||p(x) \le r$. Hence, $C_1 \subset C_2$. Let $x \notin C_1$, that is, $p(x) > r$. Since $\{x_n\}$ is dense in X there exists an x_k such that

$$p(x-x_k) < \tfrac{1}{2}(p(x) - r).$$

Thus,

$$\begin{aligned} p(x_k) &\ge p(x) - p(x-x_k) \\ &> p(x) - \tfrac{1}{2}(p(x)-r) = \tfrac{1}{2}p(x) + \tfrac{1}{2}r, \end{aligned}$$

and

$$\begin{aligned} |f_k(x) - p(x_k)| &= |f_k(x) - f_k(x_k)| \\ &\le p(x-x_k) < \tfrac{1}{2}(p(x) - r). \end{aligned}$$

Hence,

$$f_k(x) = p(x_k) - (p(x_k)-f_k(x))$$

$$\geq p(x_k) - |p(x_k)-f_k(x)|$$

$$> p(x_k) - \tfrac{1}{2}(p(x)-r)$$

$$> \tfrac{1}{2}(p(x)+r) - \tfrac{1}{2}(p(x)-r) = r,$$

that is, $x \notin C_2$. Thus, $C_1 = C_2 \in B(C)$. Since $B(C)$ is invariant under translations, $\{x: p(x-a) \leq r\}$ is also in $B(C)$. Finally, $B(X) \subset B(C)$ since $\sigma\{\{x: p(x-a) \leq r\}: a \in X, r \in R\} = B(X)$. ///

Since $f(V+Z) = f(V) + f(Z)$ is a random variable whenever V and Z are random elements in a seminormed linear space X and $f \in X^*$, the sum of two random elements in a separable seminormed linear space is a random element. However, the sum of two random elements in a nonseparable Banach space need not be a random element as the example following Lemma 2.2.1 illustrated.

In the following definition of the modes of convergence, it will be assumed that $d(V,Z)$ is a random variable when V and Z are random elements.

Definition 2.2.1 Let $\{V_n\}$ be a sequence of random elements in a semimetric space M. Then $\{V_n\}$ converges to a random element V in M

(i) with probability one or almost surely ($V_n \overset{a.s.}{\to} V$) if

$$P[\lim_{n\to\infty} d(V_n,V) = 0] = 1,$$

(ii) in probability ($V_n \overset{P}{\to} V$) if for each $\varepsilon > 0$

$$\lim_{n\to\infty} P[d(V_n,V) \geq \varepsilon] = 0, \text{ and}$$

(iii) in the rth mean, $r > 0$, ($V_n \overset{r}{\to} V$) if

$$\lim_{n\to\infty} E[d(V_n,V)^r] = 0$$

where the expected value $E[d(V_n,V)^r]$ is assumed to exist.

Other modes of convergence can be defined such as convergence in distribution [Billingsley (1968)], but they will not be discussed here since these notes will concentrate mainly on developing results using convergence with probability one and convergence in probability. Convergence results of weighted sums of random elements in which the convergence is with probability one are called strong laws, and those concerned with convergence in probability are referred to as weak laws.

When d(V,Z) is a (non-negative) random variable, the following form of Markov's inequality is valid for random elements:

$$P[d(V,Z) \geq \varepsilon] \leq \frac{E[d(V,Z)^r]}{\varepsilon^r} \tag{2.2.1}$$

for each $r > 0$ and $\varepsilon > 0$ whenever the expected value $E[d(V,Z)^r]$ exists. Inequality (2.2.1) is frequently used in obtaining both strong and weak laws for random elements in Chapters III, IV, V, VI, and VII.

Most of the relationships which exist among the various modes of convergence for random variables are also valid for random elements in semimetric spaces. The reader may find it instructive to verify some of the implications (and non-implications) of the following diagram:

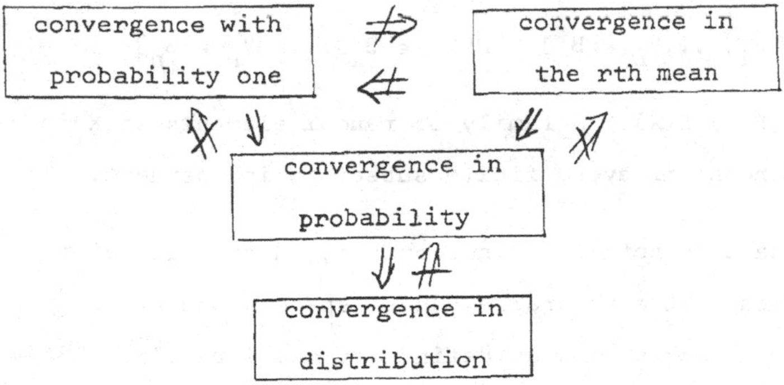

The implications are verified exactly like the random variable case, and the counterexamples can be constructed when the semimetric space is the real numbers.

2.3 PROBABILITY PROPERTIES OF RANDOM ELEMENTS

A random element in a topological space induces a probability measure on the space and its Borel subsets. In this section the definitions of identically distributed and independent random variables will be extended to random elements, and characterizations of identically distributed random elements and independent random elements will be provided for certain linear topological spaces. Finally, the Pettis integral will be used to define an expected value for random elements.

Definition 2.3.1 The random elements V and Z are said to be identically distributed if

$$P[V \in B] = P[Z \in B]$$

for each $B \in B(X)$. A family of random elements is identically distributed if every pair is identically distributed.

Definition 2.3.2 A finite set of random elements $\{V_1,\ldots,V_n\}$ in X is said to be independent if

$$P[V_1 \in B_1,\ldots,V_n \in B_n] = P[V_1 \in B_1]\ldots P[V_n \in B_n]$$

for every $B_1,\ldots,B_n \in B(X)$. A family of random elements in X is said to be independent if every finite subset is independent.

It is important to note that independent, identically distributed random elements in a separable Banach space X are strictly stationary random elements in X as defined by Mourier (1953, 1956).

Lemma 2.3.1 (a) Let $\{V_\alpha: \alpha \in A\}$ be a family of identically distributed random elements in X and let T be a Borel measurable function from X into another topological space Y. Then $\{T(V_\alpha): \alpha \in A\}$ is a family of identically distributed random elements in Y.

(b) Let $\{V_\alpha: \alpha \in A\}$ be a family of independent random elements in X and let $\{T_\alpha: \alpha \in A\}$ be a corresponding family of Borel measurable functions from X into Y. Then $\{T_\alpha(V_\alpha): \alpha \in A\}$ is a family of independent random elements in Y.

Proof: Lemma 2.1.1 provides that $T_\alpha(V_\alpha)$ is a random element in Y for each $\alpha \in A$. Let $\alpha_1,\ldots,\alpha_n \in A$ and $B_1,\ldots,B_n \in B(Y)$. Then

$$P[T_{\alpha_1}(V_{\alpha_1}) \in B_1,\ldots,T_{\alpha_n}(V_{\alpha_n}) \in B_n]$$

$$= P[V_{\alpha_1} \in T_{\alpha_1}^{-1}(B_1),\ldots,V_{\alpha_n} \in T_{\alpha_n}^{-1}(B_n)]$$

$$= P[V_{\alpha_1} \in T_{\alpha_1}^{-1}(B_1)]\ldots P[V_{\alpha_n} \in T_{\alpha_n}^{-1}(B_n)]$$

$$= P[T_{\alpha_1}(V_{\alpha_1}) \in B_1]\ldots P[T_{\alpha_n}(V_{\alpha_n}) \in B_n].$$

Thus, $\{T_\alpha(V_\alpha): \alpha \in A\}$ is a family of independent random elements in Y. The proof for identical distributions is similar. ///

Definition 2.3.3 Let C be a subfamily of B(X). If for all probability measures P and Q on B(X) P(D) = Q(D) for each $D \in C$ implies that P = Q on B(X), then C is called a family of unicity or a determining class for B(X). [See Grenander (1963), p. 128, and Billingsley (1968), p. 15.]

If C is a field and $\sigma(C) = B(X)$, then C is a family of unicity [Halmos (1950), p. 54]. Also if X is a separable seminormed linear space, then

$$C = \{\{x: f(x) < t\}: f \in X^* \text{ and } t \in R\} \qquad (2.3.1)$$

is a family of unicity [Grenander (1963), p. 128]. For the spaces $C[0,1]$ and $C[0,\infty)$, the collection

$$C = \{\{x: (x(t_1),\ldots,(t_n)) \in B\}: B \in B(R^n), \text{ and } n = 1,2,\ldots\}$$

forms a family of unicity [Billingsley (1968) and Whitt (1970)].

Lemma 2.3.2 Let X be a separable seminormed linear space. The random elements V and Z in X are identically distributed if and only if f(V) and f(Z) are identically distributed random variables for each $f \in X^*$.

Proof: Let V and Z be identically distributed random elements in X. Since each $f \in X^*$ is a continuous function from X into R, f(V) and f(Z) are identically distributed random elements in R (that is, random variables) for each $f \in X^*$ by Lemma 2.1.1.

Suppose that f(V) and f(Z) are identically distributed random variables for each $f \in X^*$. If $P_V = P_Z$ on the family of unicity which is given in (2.3.1), where P_V and P_Z are the probability measures induced on B(X) by V and Z respectively, then V and Z are identically distributed. Thus,

$$\begin{aligned} P_V[\{x: f(x) < t\}] &= P[V \in \{x: f(x) < t\}] \\ &= P[V \in f^{-1}(-\infty,t)] \\ &= P[f(V) \in (-\infty,t)] \\ &= P[f(Z) \in (-\infty,t)] \\ &= P[Z \in f^{-1}(-\infty,t)] \\ &= P[Z \in \{x: f(x) < t\}] \\ &= P_Z[\{x: f(x) < t\}]. \end{aligned}$$

Thus, V and Z are identically distributed random elements in X. ///

Lemma 2.3.3 Let X be a separable seminormed linear space. The random elements V and Z in X are independent if and only if

$f(V)$ and $g(Z)$ are independent random variables for each $f,g \in X^*$.

<u>Proof</u>: The "only if" part is immediate by Lemma 2.3.1. For the "if" part, fix $f \in X^*$ and $t \in R$. Define

$$P_{t,f}(B) = \frac{P[V \in \{x: f(x) < t\} \text{ and } Z \in B]}{P[V \in \{x: f(x) < t\}]} \qquad (2.3.2)$$

for each $B \in B(X)$. Note that $P_{t,f} = P_Z$ on $\{\{x: g(x) < s\}: g \in X^*$ and $s \in R\}$. Hence, from (2.3.1) and (2.3.2)

$$P_Z(B)P_V(\{x: f(x) < t\}) = P[V \in \{x: f(x) < t\} \text{ and } Z \in B] \quad (2.3.3)$$

for each $B \in B(X)$, $f \in X^*$, and $t \in R$. Next, for fixed $B_1 \in B(X)$ define

$$P_1(B) = \frac{P[V \in B \text{ and } Z \in B_1]}{P[Z \in B_1]} \qquad (2.3.4)$$

for each $B \in B(X)$. From (2.3.3) it follows that

$$P_1(\{x: f(x) < t\}) = \frac{P_Z[B_1]P[V \in \{x: f(x) < t\}]}{P_Z[B_1]}$$

for each $f \in X^*$ and $t \in R$, and hence,

$$P_V(B) = P_1(B) = \frac{P[V \in B \text{ and } Z \in B_1]}{P[Z \in B_1]} \qquad (2.3.5)$$

for all $B \in B(X)$ and fixed (but arbitrary) $B_1 \in B(X)$. Thus, (2.3.5) yields $P[V \in B \text{ and } Z \in B_1] = P[V \in B]P[Z \in B_1]$ for all $B,B_1 \in B(x)$.///

It is clear that both Lemma 2.3.2 and Lemma 2.3.3 extend very easily to families of random elements. However, it may not be clear that $f(V)$ and $f(Z)$ being independent for each $f \in X^*$ is not sufficient for V and Z to be independent. Let A_1 and A_2 be independent, normally distributed random variables with mean 0 and variance 1. Define random elements in $X = R^2$ by

$$V = (A_1,A_2) \qquad \text{and} \qquad Z = (A_2,-A_1). \qquad (2.3.6)$$

Each $f \in (R^2)^* = R^2$ is of the form $f(x_1,x_2) = ax_1 + bx_2$ for $a,b \in R$. Thus, $f(V) = aA_1 + bA_2$ and $f(Z) = aA_2 - bA_1$ are independent random variables [check the characteristic functions to see that $\phi_{f(V)f(Z)} = \phi_{f(V)}\phi_{g(V)}$]. But, $f_1(x_1,x_2) = x_1$ and $f_2(x_1,x_2) = x_2$ are in $(R^2)^*$, and $f_2(V) = A_2 = f_1(Z)$ which are not independent which implies that V and Z are not independent random elements.

Property 2.3.4 For random elements V and Z in the spaces s, c, c_0, ℓ^p $(p \geq 1)$, and $R^{(\infty)}$ to be identically distributed, it is necessary and sufficient that the finite-dimensional distributions be the same.

Property 2.3.4 holds since

$$C = \{\{x = (x_1,x_2,\ldots): (x_1,\ldots,x_n) \in B\}: n = 1,2,\ldots \text{ and } B \in B(R^n)\}$$

forms a field and hence a family of unicity since $\sigma(C) = B(X)$. Property 2.3.4 also holds for random elements in $C[0,1]$ and $C[0,\infty)$.

The condition that all finite-dimensional distributions of the two random elements be the same can not be reduced to requiring that the random elements be identically distributed in each coordinate. Let $\Omega = \{H,T\}$ with $A = 2^\Omega$ and $P(\{H\}) = \frac{1}{2} = P(\{T\})$. Let $A_1(H) = 1 = A_2(T)$ and $A_1(T) = -1 = A_2(H)$. Define $V = (A_1,A_2,0,0,\ldots)$ and $Z = (A_1,A_1,0,0,\ldots)$. The random elements V and Z (in s, c, c_0, ℓ^p, or $R^{(\infty)}$) are identically distributed in each coordinate. But, $f(x) = x_1 + x_2$ is a continuous linear functional, and $f(V) = A_1 + A_2 = 0$ is not distributed as $f(Z) = 2A_1$. Thus, V and Z are not identically distributed by Lemma 2.3.1. To extend this counter-example to $C[0,1]$, let

$$x(t) = \begin{cases} 1-2t & \text{if } 0 \leq t \leq \frac{1}{2} \\ 0 & \text{if } \frac{1}{2} < t \leq 1 \end{cases}$$

and

$$y(t) = \begin{cases} 0 & \text{if } 0 \le t \le \frac{1}{2} \\ 2t-1 & \text{if } \frac{1}{2} < t \le 1. \end{cases}$$

Then define $V(t) = A_1x(t) + A_2y(t)$ and $Z(t) = A_1x(t) + A_1y(t)$. Note that for each $t \in [0,1]$, $V(t)$ and $Z(t)$ are identically distributed. But the random elements V and Z in C[0,1] are not identically distributed since $f(V) = V(\frac{1}{4}) + V(\frac{3}{4}) = \frac{1}{2}A_1 + \frac{1}{2}A_2 = 0$ and $f(Z) = Z(\frac{1}{4}) + V(\frac{1}{4}) = \frac{1}{2}A_1 + \frac{1}{2}A_1 = A_1$ where $f(x) = x(\frac{1}{4}) + x(\frac{3}{4}) \in C[0,1]^*$.

For stochastic processes which have sample paths in C[0,1], C[0,∞), or D[0,1], an increment condition was shown to be necessary and sufficient for identical distributions by Taylor (1976).

<u>Property 2.3.5</u> Let V(t) and Z(t) be stochastic processes such that $P[V(0) = 0] = 1 = P[Z(0) = 0]$ which have independent increments and whose sample paths are in C[0,1], C[0,∞), or D[0,1] (see Chapter VII) with probability one. Then V and Z are identically distributed random elements if and only if $V(t) - V(s)$ is distributed as $Z(t) - Z(s)$ for each t and s.

<u>Proof</u>: The necessity is immediate since $f(x) = x(t) - x(s)$ is a Borel measurable functional. The sufficiency will be obtained by showing that the characteristic functions of the finite-dimensional distributions agree.

For $0 < t_1 < \ldots < t_n$, observe that

$$E[\exp(i[a_1Z(t_1) + \ldots + a_nZ(t_n)])]$$

$$= E[\exp(i(a_n + \ldots + a_1)(Z(t_1)-Z(0)) + \ldots + ia_n(Z(t_n)-Z(t_{n-1})))]$$

$$= E[\exp(i(a_n + \ldots + a_1)(Z(t_1)-Z(0))]\ldots E[\exp(ia_n\ (Z(t_n)-Z(t_{n-1})))]$$

$$= E[\exp(i(a_n + \ldots + a_1)(V(t_1)-V(0))]\ldots E[\exp(ia_n\ (V(t_n)-V(t_{n-1})))]$$

$$= E[\exp(i[a_1V(t_1) + \ldots + a_nV(t_n)])] \qquad (2.3.7)$$

for all $a_1,\ldots,a_n \in R$. Hence, from (2.3.7) it follows that V and Z are identically distributed. ///

If in addition to the hypotheses in Property 2.3.5 the stochastic process has stationary increments then the condition that V(t) is distributed as Z(t) for each t becomes necessary and sufficient for identical distributions. Thus, Property 2.3.5 shows that two Brownian motion processes V and Z are identically distributed if and only if they have the same parameter $\sigma_V^2 = E[V(1)^2] = E[Z(1)^2] = \sigma_Z^2$. Similar results hold for Poisson processes when viewed as random elements in D[0,1] (see Chapter VII). Finally, it is clear that Property 2.3.5 will hold for any subspace of $R^{[0,\infty)}$ where the cylinder sets form a family of unicity for the Borel subsets of the subspace.

Lemma 2.3.3 provided a characterization of independence in terms of the dual space. When X is a linear metric space with a Schauder basis $\{b_k\}$, then a characterization of independence is available in terms of the coordinate functionals $\{f_k\}$. For notational convenience the result is only listed for pairwise independence.

<u>Property 2.3.6</u> Let X be a normed linear space which has a Schauder basis $\{b_k\}$ and Borel measurable coordinate functionals $\{f_k\}$. The random elements V and Z in X are independent if and only if the random vectors $(f_1(V),\ldots,f_n(V))$ and $(f_1(Z),\ldots,f_n(Z))$ are independent for each $n = 1,2,\ldots$.

<u>Proof</u>: The "only if" part is immediate since $f(x) = (f_1(x),\ldots,f_n(x))$ is a Borel measurable function from X to R^n. Next, assume that $(f_1(V),\ldots,f_n(V))$ is independent of $(f_1(Z),\ldots,f_n(Z))$ for each n. For each $f,g \in X^*$ and for each n,

$$f(U_n(V)) = f(\textstyle\sum_{k=1}^n f_k(V)b_k) = \sum_{k=1}^n f_k(V)f(b_k)$$

is independent of

$$g(U_n(Z)) = \sum_{k=1}^{n} f_k(Z)g(b_k).$$

Since $\lim_{n\to\infty} U_n(x) = x$ for each $x \in X$, then

$$\lim_{n\to\infty} f(U_n(V)) = f(V) \text{ and } \lim_{n\to\infty} g(U_n(Z)) = g(Z)$$

pointwise. Hence,

$$\begin{aligned}
&E[\exp(itf(V) + isg(Z))] \\
&= E[\lim_{n\to\infty} \exp(itf(U_n(V)) + isg(U_n(Z)))] \\
&= \lim_{n\to\infty} E[\exp(itf(U_n(V)) + isg(U_n(Z)))] \\
&= \lim_{n\to\infty} E[\exp(itf(U_n(V)))] \lim_{n\to\infty} E[\exp(isg(U_n(Z)))] \\
&= E[\lim_{n\to\infty} \exp(itf(U_n(V)))] E[\lim_{n\to\infty} \exp(isg(U_n(Z)))] \\
&= E[\exp(itf(V))] E[\exp(isg(Z))] \qquad (2.3.8)
\end{aligned}$$

for each $t,s \in R$. From Equation (2.3.8) the random variables $f(V)$ and $g(Z)$ are independent for each $f,g \in X^*$. Thus, the random elements V and Z are independent by Lemma 2.3.3. ///

Property 2.3.6 is also valid for the Fréchet spaces s and $C[0,\infty)$. A lemma similar to Lemma 2.3.3 holds for s and $C[0,\infty)$, and the proof of Property 2.3.6 then follows. Thus, random elements in the sequence spaces, $C[0,\infty)$, and $C[0,1]$ are independent if and only if the random vectors of their respective finite-dimensional distributions are independent. It is also easy to see that pointwise independence or coordinate independence is not sufficient for independence of the random elements. The examples similar to those showing that coordinate identical distributions did not suffice can be constructed.

An expected value of a random element in a linear topological space can be defined by a Pettis integral [Pettis (1938)].

<u>Definition 2.3.4</u> A random element V in a linear topological space X is said to have <u>expected value</u> EV if there exists an element EV $\in$ X such that $E[f(V)] = f(EV)$ for each $f \in X^*$.

If X^* is separating over X [see Wilansky (1964)], then the expected value EV is unique. If $EV \neq x_0$, then there is $f \in X^*$ such that $E[f(V)] = f(EV) \neq f(x_0)$. Thus, $f(x_0)$ can not equal $E[f(V)]$ for every $f \in X^*$ unless $x_0 = EV$.

<u>Definition 2.3.5</u> Let X be a seminormed linear space with seminorm p and let V be a random element in X with expected value EV. Then the <u>variance</u> of V can be defined by

$$\sigma_V^2 = E[p(V - EV)^2],$$

and the <u>standard deviation</u> of V, σ_V, is the non-negative square root of the variance.

The properties of the expected value which are listed in Theorem 2.3.7 below are consequences of the definition of the Pettis integral (a separating dual space will be assumed to avoid ambiguity).

<u>Theorem 2.3.7</u> Let V, V_1, and V_2 be random elements in a linear topological space X and let $x \in X$.

(i) If EV_1 and EV_2 exists and $V_1 + V_2$ is a random element in X, then $E[V_1 + V_2] = EV_1 + EV_2$.

(ii) If EV exists and $b \in R$, then $E(bV) = bEV$.

(iii) If $V = x$ with probability one and A is a random variable such that EA exists, then $E(AV) = (EA)x$.

(iv) If h is a continuous linear function from X into a linear topological space Y and if EV exists, then $E[hV] = h(EV)$.

(v) If EV exists and X is a seminormed linear space with seminorm p, then $p(EV) \leq E[p(V)]$ where $E[p(V)]$ may be infinite.

Proof: Properties (i), (ii), & (iii) follow immediately from the definition EV and the linearity of $f \in X^*$.

For (iv) note that $g \circ h \in X^*$ for each $g \in Y^*$. Thus,

$$g(h(EV)) = (g \circ h)(EV) = E[(g \circ h)(V)] = E[g(hV)].$$

Hence, $E[hV] = h(EV)$ and $h(EV) \in Y$.

To prove part (v), notice that if $EV = 0$ (the zero element) then the statement is true. If $EV \neq 0$ then by Corollary 1.2.3 there exists an $f \in X^*$ such that $||f|| = 1$ and $f(EV) = p(EV)$. Hence,

$$\begin{aligned} p(EV) = |f(EV)| &= |Ef(V)| \\ &\leq E|f(V)| \\ &\leq E[||f||p(V)] \\ &= E[p(V)]. \end{aligned}$$ ///

A Bochner integral is defined for a countably-valued strongly measurable random element V in a Banach space X by

$$(B) \int_\Omega V dP = \sum_{k=1}^\infty x_k P[V = x_k] \tag{2.3.9}$$

when $\sum_{k=1}^\infty ||x_k||P[V = x_k] = \int_\Omega ||V||dP < \infty$. An arbitrary strongly measurable random element in a Banach space X is said to be Bochner integrable if and only if there exists a sequence of countably-valued, Bochner integrable random elements $\{V_n\}$ converging almost surely to V and such that

$$\lim_{n\to\infty} \int_\Omega ||V - V_n||dP = 0. \tag{2.3.10}$$

In which case,

$$(B)\int_\Omega V\,dP = \lim_{n\to\infty} (B)\int_\Omega V_n\,dP. \qquad (2.3.11)$$

A necessary and sufficient condition for the Bochner integral to exist for a strongly measurable random element V in a Banach space is that

$$\int_\Omega ||V||dP < \infty \qquad (2.3.12)$$

(see Hille and Phillips (1957), pages 78-92). The following result gives a sufficient condition for the expected value (the Pettis integral) to exist and illustrates some relationships between the Bochner integral and the Pettis integral.

<u>Lemma 2.3.8</u> Let X be a separable Banach space and let V be a random element in X. If $E||V|| < \infty$, then EV exists.

<u>Proof</u>: If V is countably-valued and takes values $\{a_i\} \subset X$, then $E||V|| < \infty$ implies that the series

$$\sum_{i=1}^{\infty} a_i P[V = a_i] \qquad (2.3.13)$$

converges absolutely. Since X is complete, the series converges to some $EV \in X$. For $f \in X^*$

$$\begin{aligned} f(EV) &= f(\textstyle\sum_{i=1}^{\infty} a_i P[V = a_i]) \\ &= f(\lim_{n\to\infty} \textstyle\sum_{i=1}^{n} a_i P[V = a_i]) \\ &= \lim_{n\to\infty} \textstyle\sum_{i=1}^{n} f(a_i) P[V = a_i] \\ &= \textstyle\sum_{i=1}^{\infty} f(a_i) P[V = a_i] = E[f(V)]. \end{aligned} \qquad (2.3.14)$$

Let V be an arbitrary random element in X. By Property 2.1.4 there exists a sequence of countably-valued random elements $\{V_n\}$ which converge uniformly to V. Hence, EV_n exists for each n. For any n and m

$$||EV_n - EV_m|| = ||E(V_n - V_m)|| \leq E||V_n - V_m|| \qquad (2.3.15)$$

by Theorem 2.3.7 (i) and (v). Since $V_n \to V$ uniformly, $||EV_n - EV_m||$ can be made arbitrarily small for m and n sufficiently large. Therefore, $\{EV_n\}$ is a Cauchy sequence in the complete space X and converges to a limit, say $EV \in X$. For each $f \in X^*$

$$\begin{aligned} f(EV) &= f(\lim_{n\to\infty} EV_n) \\ &= \lim_{n\to\infty} f(EV_n) \\ &= \lim_{n\to\infty} E[f(V_n)] \\ &= E[\lim_{n\to\infty} f(V_n)] \qquad \text{(dominated convergence theorem)} \\ &= E[f(V)] \qquad (||V_n - V|| \to 0 \text{ uniformly}) \end{aligned}$$

by the continuity of f. Hence, EV is the expected value of V. ///

While the norm could be replaced by a seminorm, the hypothesis of completeness is essential in Lemma 2.3.8. For example, consider the normed linear space $R^{(\infty)}$ (rf. Section 1.3, (3)) with $V = \delta^n$ with probability $1/2^n$ where $\delta^1 = (1,0,0,\ldots)$, $\delta^2 = (0,1,0,\ldots),\ldots$. Then

$$E||V|| = \sum_{n=1}^{\infty} \frac{1}{2^n} = 1$$

which is $< \infty$ but if EV existed it would be of the form

$$EV = (\tfrac{1}{2}, \tfrac{1}{4}, \tfrac{1}{8}, \ldots, \tfrac{1}{2^n}, \ldots)$$

which is not an element of $R^{(\infty)}$.

For a different example, let X be the separable Hilbert space ℓ^2. Define $V = n\delta^n$ with probability c/n^2 where c is the appropriate constant. Note that

$$E||V|| = \sum_{n=1}^{\infty} n \frac{c}{n^2} = \infty$$

but that

$$EV = (\frac{c}{1}, \frac{c}{2}, \frac{c}{3}, \ldots, \frac{c}{n}, \ldots) \in \ell^2.$$

Thus, EV exists but $E||V|| = \infty$.

Let X be a linear topological space which has a Schauder basis $\{b_k\}$ such that the coordinate functionals $\{f_k\}$ are continuous. If V is a random element in X which has expected value $EV \in X$, then $f_k(EV) = E[f_k(V)]$ since each $f_k \in X^*$. Thus,

$$\begin{aligned} EV &= \sum_{k=1}^{\infty} f_k(EV) b_k \\ &= \sum_{k=1}^{\infty} E[f_k(V)] b_k. \end{aligned} \tag{2.3.16}$$

In the sequence spaces, Equation (2.3.16) provides that

$$EV = (EV_1, EV_2, \ldots)$$

where $V = (V_1, V_2, \ldots)$. Thus, for stochastic processes with countable parameter space, the (Pettis integral) expected value agrees with the mean value function. For the space $C[0,1]$ or $C[0,\infty)$, $f_t(x) = x(t)$ defines a continuous linear functional. Thus, $f_t(EV) = E[f_t(V)] = E(V(t))$, or EV is the pointwise expected value of the stochastic process.

2.4 PROBLEMS

2.1 Show that $V^{-1}(C) \in A$ for each closed set C in X implies that $V^{-1}(E) \in A$ for each $E \in B(X)$.

2.2 In Equation (2.1.1) of Section 2.1, verify that

$$V^{-1}(C) = \bigcap_{k=1}^{\infty} \bigcup_{n=1}^{\infty} \bigcap_{m=n}^{\infty} V_m^{-1}(C_k).$$

2.3 Let $\{V_n\}$ be a sequence of random variables. Show that $V = (V_1, V_2, \ldots)$ is a random element in s. [Hint: use technique in (2.1.3).]

2.4 Show that the functions V and Z which were defined in (2.2.1) are random elements in X. [Recall that $B(X) \times B(X)$ is the sigma field generated by sets of the form $B_1 \times B_2$ where $B_1, B_2 \in B(X)$.]

2.5 Verify the form of Markov's inequality which is listed in Inequality (2.2.1).

2.6 Prove or disprove. If $\{V_\alpha: \alpha \in A\}$ is a family of identically distributed random elements in a linear topological space X and $\{T_\alpha: \alpha \in A\}$ is a corresponding family of Borel measurable functions from X into the real numbers R, then $\{T_\alpha(V_\alpha): \alpha \in A\}$ is a family of identically distributed random variables.

2.7 Show that the random variables f(V) and f(Z) are independent where $f(x_1,x_2) = ax_1 + bx_2$ with $a,b \in R$ and V and Z are the random elements which are defined in (2.3.6).

2.8 Construct random elements V and Z in C[0,1] such that V(t) and Z(t) are independent random variables for each $t \in [0,1]$ but such that V and Z are not independent random elements.

2.9 Show that random elements in the sequence spaces s, c, c_0, and ℓ^p $(p \geq 1)$ can be independent in each coordinate without being independent random elements.

CHAPTER III

LAWS OF LARGE NUMBERS, UNCORRELATION, AND CONVERGENCE OF WEIGHTED SUMS OF RANDOM VARIABLES

3.0 INTRODUCTION

This chapter will contain some of the standard laws of large numbers and some weighted sum convergence results for random variables. Also, several laws of large numbers for random elements in separable Hilbert spaces are presented. Essentially, the first and fourth sections will consist of material which will be used for reference in proving the stochastic convergence results in Chapters IV, V, VI and VII. Thus, not all of the existing results for random variables will be given here.

In Section 3.2 several of the laws of large numbers will be extended to random elements in separable Hilbert spaces. The key points in these extensions will be the use of the inner product to define uncorrelated random elements and the fact that the variance of a sum of uncorrelated random elements in the sum of the variances for random elements in separable Hilbert spaces.

In Section 3.3 the concept of uncorrelation and the several possible definitions of uncorrelation are examined. The comparison of restrictions on the random elements will be valuable in contrasting the various weak laws of large numbers. Finally, Section 3.4 will list several convergence results for weighted sums of random variables.

3.1 LAWS OF LARGE NUMBERS FOR RANDOM VARIABLES

A review of some of the existing laws of large numbers will be presented in this section. These will be used primarily as background material for the extensions of the laws of large numbers to separable Hilbert spaces, separable normed linear spaces, and certain Fréchet spaces. Several very general laws of large numbers and their proofs are contained in standard textbooks on probability theory [Chung (1968), Doob (1953), Loéve (1963), Neveu (1965), Tucker (1967), and others] and hence will not be reproduced in detail here.

Definition 3.1.1 Two random variables V and Z with $E(V^2) < \infty$ and $E(Z^2) < \infty$ are said to be uncorrelated if their covariance is zero, that is, if $\text{Cov}(V,Z) = E[(V-EZ)(Z-EZ)] = 0$. A family of random variables is said to be uncorrelated if every pair is uncorrelated.

Since random variables are random elements, the definitions of the various modes of convergence for random variables and for independent and identically distributed random variables were given in Chapter II.

Theorem 3.1.1 If $\{V_n\}$ is a sequence of uncorrelated random variables such that

$$(\frac{1}{n})^2 \sum_{k=1}^{n} \text{Var}(V_k) \to 0$$

as $n \to \infty$, then

$$\frac{1}{n} \sum_{k=1}^{n} (V_k - EV_k) \to 0$$

in probability.

Theorem 3.1.1 is a special case of Theorem 3.2.1 which will give a weak law of large numbers for separable Hilbert spaces. Thus, the proof of Theorem 3.1.1 can be obtained from the proof of Theorem 3.2.1 by taking the separable Hilbert space to be the real numbers.

An immediate corollary to Theorem 3.1.1 is the weak law of large numbers for identically distributed random variables which are uncorrelated since

$$(\frac{1}{n})^2 \sum_{k=1}^{n} \operatorname{Var}(V_k) = \frac{1}{n} \operatorname{Var}(V_1) \to 0$$

as $n \to \infty$. However, a more interesting result follows with the less restrictive condition of uniformly bounded variances of the uncorrelated random variables. In this case the weak and strong laws of large numbers both hold.

Thoerem 3.1.2 If $\{V_n\}$ is a sequence of uncorrelated random variables such that for all n, $\operatorname{Var}(V_n) \le M$ where M is a constant, then

$$\frac{1}{n} \sum_{k=1}^{n} (V_k - EV_k) \to 0$$

with probability one.

Outline of Proof [For more details see Chung (1968), page 97] For each n let

$$S_n = \sum_{k=1}^{n} (V_k - EV_k).$$

For each $\varepsilon > 0$

$$\sum_{n=1}^{\infty} P[|S_{n^2}| > n^2\varepsilon] \le \sum_{n=1}^{\infty} \frac{M}{n^2\varepsilon^2} < \varepsilon$$

implies that $P[|S_{n^2}| > n^2\varepsilon \text{ i.o.}] = 0$. Next

$$P[\max_{n^2 \le k < (n+1)^2} |S_k - S_{n^2}| > n^2 \varepsilon] \le \frac{4M}{\varepsilon^2 n^2}$$

implies that

$$P[\max_{n^2 \le k < (n+1)^2} |S_k - S_{n^2}| > n^2 \varepsilon \text{ i.o.}] = 0.$$

Hence,

$$\frac{|S_k|}{k} \le \frac{|S_{n^2}| + \max_{n^2 \le k < (n+1)^2} |S_k - S_{n^2}|}{n^2} \to 0$$

with probability one. ///

An even stronger result is given by the following theorem [Doob (1953), page 158].

Theorem 3.1.3 If $\{V_n\}$ is a sequence of uncorrelated random variables such that

$$\sum_{n=1}^{\infty} \frac{\text{Var}(V_n)}{n^2} \log^2 n < \infty, \tag{3.1.1}$$

then

$$\frac{1}{n} \sum_{k=1}^{n} (V_k - EV_k) \to 0$$

with probability one.

The proof of Theorem 3.1.3 is similar to the proof of Theorem 3.1.2 with the maximum being taken from n to 2n and with different constants in the sums. In addition, this result easily extends to separable Hilbert spaces by following the proof in Doob (1953) using uncorrelated random elements which will be defined in Section 3.2 of the present chapter.

For independent random variables a strong law of large numbers holds under a weaker condition than (3.1.1). Although more general results hold [see Chung (1947) or Chung (1968), page 117], the following statement of Kolmogorov's strong law of large numbers for non-identically distributed random variables will be needed in proving laws of large numbers in linear topological spaces.

Theorem 3.1.4 If $\{V_n\}$ is a sequence of independent random variables such that

$$\sum_{n=1}^{\infty} \frac{\mathrm{Var}(V_n)}{n^2} < \infty, \tag{3.1.2}$$

then

$$\frac{1}{n} \sum_{k=1}^{n} (V_k - EV_k) \to 0$$

with probability one.

Outline of Proof [For more details see Tucker (1967), p. 124]: Condition (3.1.2) implies that

$$\sum_{n=1}^{\infty} (V_n - EV_n)/n$$

converges almost surely. Thus, by Kronecker's Lemma

$$\frac{1}{n} \sum_{k=1}^{n} (V_k - EV_k) \to 0$$

with probability one. ///

The last law of large numbers for random variables to be listed is Kolmogorov's strong law for independent, identically distributed random variables which requires only that the first moments exist.

Theorem 3.1.5 If V_n is a sequence of independent, identically distributed random variables such that $E|V_1| < \infty$, then

$$\frac{1}{n} \sum_{k=1}^{n} V_k \to EV_1$$

with probability one. Moreover, the converse is true.

The proof of Theorem 3.1.5 is obtained by truncation, the three-series theorem, and Theorem 3.1.4. For the converse, the Borel lemmas are used. Additional details on the proofs of these laws of large numbers are available in Chapter 5 of Chung (1968), Chapter 5 of Loéve (1963), and Chapter 5 of Tucker (1967).

3.2 EXTENSIONS TO SEPARABLE HILBERT SPACES

In this section the laws of large numbers which were listed in Section 3.1 will be extended to separable Hilbert spaces. Independent random elements were defined in Section 2.3, and some related properties were given. Uncorrelated random elements in separable Hilbert spaces will be defined by using the inner product. It will be shown that in a separable Hilbert space the variance of a sum of uncorrelated random elements is the sum of their variances and that independent random elements are uncorrelated. Once this is accomplished, the extensions of the laws of large numbers to separable Hilbert spaces will follow from the proofs of the corresponding laws of large numbers for random variables.

In attempting to extend weak laws of large numbers for uncorrelated random variables to linear topological spaces, one major problem is defining uncorrelated random elements since multiplication of elements is not generally defined. However, a definition of multiplication in a Hilbert space is given in terms of an inner product, and uncorrelated random elements can be defined by the expectation of the inner product of the random elements.

Let V and Z be two random elements in a Hilbert space H with $E||V||^2 < \infty$ and $E||Z||^2 < \infty$. If H is separable, then Lemma 2.2.2 implies that V + Z and V - Z are random elements. Thus,

$$\langle V,Z\rangle = \frac{1}{4}\left(||V + Z||^2 - ||V - Z||^2\right)$$

is a random variable since $||V + Z||$ and $||V - Z||$ are random variables. [For $x,y \in H$, then $\langle x,y\rangle = \frac{1}{4}(||x+y||^2 - ||x-y||^2)$ by the parallelogram law, Wilansky (1964), pages 124-125.] Moreover, $E\langle V,Z\rangle$ is defined since

$$E|\langle V,Z\rangle| \leq E(||V||\ ||Z||) \leq (E||V||^2)^{\frac{1}{2}}(E||Z||^2)^{\frac{1}{2}} < \infty$$

by the Cauchy-Schwarz-Bunyakowski inequality. Also, $E||V||^2 < \infty$ implies that $E||V|| < \infty$ and that EV exists by Lemma 2.3.8 since H is separable and complete.

Definition 3.2.1 Two random elements V and Z in a separable Hilbert space with $E||V||^2 < \infty$ and $E||Z||^2 < \infty$ are said to be uncorrelated if $E\langle V,Z\rangle = \langle EV,EZ\rangle$. A family of random elements is said to be uncorrelated if every pair of random elements in the family is uncorrelated.

Definition 3.2.1 is a direct generalization of the concept of uncorrelated random variables (see Definition 3.1.1). Furthermore, the covariance of the random elements V and Z can be defined as

$$\mathrm{Cov}\ (V,Z) = E\langle V-EV,\ Z-EZ\rangle. \tag{3.2.1}$$

For each $x \in H$, $f_x(y) = \langle x,y\rangle = \langle y,x\rangle$ is a continuous linear functional. Since $EV \in H$ and $EZ \in H$,

$$\begin{aligned} E\langle V-EV,Z-EZ\rangle &= E\langle V,Z\rangle - E\langle EV,Z\rangle - E\langle V,EZ\rangle + \langle EV,EZ\rangle \\ &= E\langle V,Z\rangle - \langle EV,EZ\rangle - \langle EV,EZ\rangle + \langle EV,EZ\rangle \\ &= E\langle V,Z\rangle - \langle EV,EZ\rangle. \end{aligned}$$

Thus, (3.2.1) can also be expressed as

$$\mathrm{Cov}\ (V,Z) = E\langle V,Z\rangle - \langle EV,EZ\rangle. \tag{3.2.2}$$

The variance of a random element V in a separable Hilbert space naturally follows as

$$\text{Var}\ (V) = E||V-EV||^2 = E<V-EV,V-EV>. \qquad (3.2.3)$$

From this definition it follows that the variance of a finite sum of uncorrelated random elements in a separable Hilbert space is the sum of the variances. Thus, the laws of large numbers which were given in Section 3.1 generalize immediately to separable Hilbert spaces (see Theorems 3.2.1, 3.2.2, and 3.2.4 below). However, even though it is very useful and intuitive, Definition 3.2.1 has surprising implications even for finite-dimensional spaces. This will be illustrated by Example 3.3.1 in which two random elements in R^2 will be shown to be uncorrelated even though they agree in the first coordinate and differ only in sign in the second coordinate.

Theorem 3.2.1 If $\{V_n\}$ is a sequence of uncorrelated random elements in a separable Hilbert space such that

$$(\frac{1}{n})^2 \sum_{k=1}^n \text{Var}\ (V_k) \to 0 \qquad (3.2.3)$$

as $n \to \infty$, then

$$||\frac{1}{n} \sum_{k=1}^n (V_k-EV_k)|| \to 0$$

in probability.

Proof: Let $\varepsilon > 0$ be given. By the Markov inequality (2.2.1)

$$P[||\frac{1}{n} \sum_{k=1}^n (V_k-EV_k)|| > \varepsilon] \le (\frac{1}{\varepsilon n})^2 E||\sum_{k=1}^n (V_k-EV_k)||^2$$

$$= (\frac{1}{\varepsilon n})^2 E<\sum_{k=1}^n (V_k-EV_k), \sum_{j=1}^n (V_j-EV_j)>$$

$$= (\frac{1}{\varepsilon n})^2 \sum_{k=1}^n \sum_{j=1}^n E<V_k-EV_k,V_j-EV_j>

$$= \left(\frac{1}{\varepsilon n}\right)^2 \sum_{k=1}^{n} E\langle V_k - EV_k, V_k - EV_k\rangle$$

$$= \left(\frac{1}{\varepsilon}\right)^2 \left(\frac{1}{n}\right)^2 \sum_{k=1}^{n} E||V_k - EV_k||^2 \qquad (3.2.4)$$

which goes to zero as $n \to \infty$. ///

The proof of Theorem 3.2.1 is analagous to the proof of the weak law of large numbers for random variables (Thoerem 3.1.1) except for the verification of (3.2.4) which shows that the variance of a sum of uncorrelated random elements is the sum of their variances. The extension of the strong law of large numbers for uncorrelated random variables (Theorem 3.1.3) to separable Hilbert spaces follows similarly [Beck and Warren (1968)].

<u>Theorem 3.2.2</u> If $\{V_n\}$ is a sequence of uncorrelated random elements in a separable Hilbert space such that

$$\sum_{n=1}^{\infty} \frac{\text{Var }(V_n)}{n^2} \log^2 n < \infty,$$

then

$$||\frac{1}{n} \sum_{k=1}^{n} (V_k - EV_k)|| \to 0$$

with probability one.

Before presenting a strong law of large numbers for independent random elements in separable Hilbert spaces, it is important to observe that independent random elements are uncorrelated random elements when the second absolute moments exists.

<u>Lemma 3.2.3</u> If V and Z are independent random elements in a separable Hilbert space H with $E||V||^2 < \infty$ and $E||Z||^2 < \infty$, then V and Z are uncorrelated.

<u>Proof</u>: Without loss of generality, it may be assumed that $EV = 0 = EZ$; otherwise, consider the random elements V-EV and Z-EZ.

Since H is a separable Hilbert space, there exists an orthonormal basis $\{b_n\}$ [Wilansky (1964), page 130]. Moreover, by Remark 1.3.1

$$\langle V,Z\rangle = \sum_{n=1}^{\infty} \langle V,b_n\rangle\langle Z,b_n\rangle. \tag{3.2.5}$$

Furthermore, for each positive integer m

$$\begin{aligned} \left|\sum_{n=1}^{m} \langle V,b_n\rangle\langle Z,b_n\rangle\right| &\le \sum_{n=1}^{m} |\langle V,b_n\rangle\langle Z,b_n\rangle| \\ &\le \left(\sum_{n=1}^{m} |\langle V,b_n\rangle|^2\right)^{\frac{1}{2}}\left(\sum_{n=1}^{m} |\langle Z,b_n\rangle|^2\right)^{\frac{1}{2}} \\ &\le ||V||\ ||Z|| \end{aligned} \tag{3.2.6}$$

by the inequalities of Hölder and Bessel. Since

$$E(||V||\ ||Z||) \le (E||V||^2)^{\frac{1}{2}}(E||Z||^2)^{\frac{1}{2}} < \infty,$$

(3.2.5, (3.2.6), and the Lebesgue dominated convergence theorem imply that

$$\begin{aligned} E\langle V,Z\rangle &= E(\lim_{m\to\infty} \sum_{n=1}^{m} \langle V,b_n\rangle\langle Z,b_n\rangle) \\ &= \lim_{m\to\infty} \sum_{n=1}^{m} E(\langle V,b_n\rangle\langle Z,b_n\rangle). \end{aligned} \tag{3.2.7}$$

But, $b_n \in H^*$, and $\langle V,b_n\rangle$ and $\langle Z,b_n\rangle$ are independent random variables by Lemma 2.3.3. Thus

$$\begin{aligned} E(\langle V,b_n\rangle\langle Z,b_n\rangle) &= E\langle V,b_n\rangle E\langle Z,b_n\rangle \\ &= \langle EV,b_n\rangle\langle EZ,b_n\rangle \\ &= \langle 0,b_n\rangle\langle 0,b_n\rangle = 0 \end{aligned}$$

for each n. Hence, from Equation (3.2.7)

$$E\langle Z,V\rangle = \lim_{m\to\infty} \sum_{n=1}^{m} 0 = 0.$$

Thus the random elements V and Z are uncorrelated. ///

Using Lemma 3.2.3 and the equality of (3.2.4), then Kolmogorov's inequalities and the related convergence of series of independent random elements in separable Hilbert spaces can be obtained. Thus, by an extension of Kronecker's lemma for elements in a Hilbert space, the following strong law of large numbers can be obtained in an analagous manner to the random variable case.

Theorem 3.2.4 If $\{V_n\}$ is a sequence of independent random elements in a separable Hilbert space such that

$$\sum_{n=1}^{\infty} \frac{\mathrm{Var}\,(V_n)}{n^2} < \infty,$$

then

$$\left|\left|\frac{1}{n} \sum_{k=1}^{n} (V_k - EV_k)\right|\right| \to 0$$

with probability one.

The strong law of large numbers for independent, identically distributed random varaibles (Theorem 3.1.5) can similarly be extended to separable Hilbert spaces. However, this result can also be proved for separable normed linear spaces and hence will be presented (with its proof) in Section 4.1. Unfortunately, Theorem 3.1.5 is the only law of large numbers from Section 3.1 which will extend directly to separable normed linear spaces. In order to extend the other laws of large numbers to normed linear spaces, additional restrictions on either the spaces or the random elements are needed. These results are described in more detail in Chapter IV.

3.3 COMPARISONS OF UNCORRELATION CONCEPTS

In this section a comparison of the different concepts of uncorrelation will be considered for random elements. Uncorrelated random elements in a separable Hilbert spaces were obtained in

Definition 3.2.1 by using the inner product. Coordinate uncorrelated random elements can be defined when the linear topological space has a Schauder basis. The most general definition of uncorrelation will be defined by using the dual space of the linear topological space and will be called weakly uncorrelated random elements.

<u>Definition 3.3.1</u> Let X be a linear topological space and let V and Z be random elements in X such that $E[f(V)^2] < \infty$ and $E[f(Z)^2] < \infty$ for each $f \in X^*$. If

$$\mathrm{Cov}\,(f(V), f(Z)) = 0$$

for each $f \in X^*$, then V and Z are said to be <u>weakly</u> <u>uncorrelated</u> random elements in X. A family of random elements in X is said to be weakly uncorrelated if every pair is weakly uncorrelated.

<u>Definition 3.3.2</u> Let X be linear topological space which has a Schauder basis $\{b_k\}$ and coordinate functionals $\{f_k\}$. Let V and Z be random elements in X such that $f_k(V)$ and $f_k(Z)$ are random variables with finite second moments for each k. If

$$\mathrm{Cov}\,(f_k(V),\ f_k(Z)) = 0$$

for each k, then V and Z are said to be <u>coordinate</u> <u>uncorrelated</u> random elements (with respect to the basis $\{b_k\}$). A family of random elements is said to be coordinate uncorrelated if every pair is coordinate uncorrelated.

From Section 2.3 it is clear that independent random elements with finite second absolute moments will imply weakly uncorrelation. Also, it easily follows that weakly uncorrelated random elements are coordinate uncorrelated random elements for each Schauder basis which has continuous coordinate functionals. A necessary and sufficient condition for weakly uncorrelated random elements will be given in terms of the coordinate functionals of a Schauder basis. Also, it will be shown that random elements in a separable Hilbert space which are coordinate uncorrelated with respect to an orthonormal basis are uncorrelated. Then it will follow that weakly uncorrelated random elements in a separable Hilbert space are uncorrelated. However, it will be shown that no implications need necessarily exist between coordinate uncorrelated random elements and uncorrelated random elements.

Theorem 3.3.1 Let X be a Banach space which has a Schauder basis $\{b_n\}$ and coordinate functionals $\{f_n\}$. Let V and Z be random elements in X such that $E||V||^2 < \infty$ and $E||Z||^2 < \infty$. The random elements V and Z are weakly uncorrelated if and only if

$$\mathrm{Cov}\,(f_n(V) + f_k(V),\, f_n(Z) + f_k(Z)) = 0 \qquad (3.3.1)$$

for each k and n.

Proof: Obviously, (3.3.1) is necessary since $f_n + f_k \in X^*$. For the converse it can be assumed that $EV = 0 = EZ$ without loss of generality [EV and EZ exist since X is complete and separable and since $E||V|| < \infty$ and $E||Z|| < \infty$]. Thus, for any $g \in X^*$ it must be shown that

$$E[g(V)g(Z)] = 0. \qquad (3.3.2)$$

For each n define $g_n = g \circ U_n$. Thus, $g_n \in X^*$, and

$$|g_n(x)| \le ||g||\ ||U_n(x)|| \le m||g||\ ||x|| \qquad (3.3.3)$$

for each $x \in X$ and each n by Lemma 1.3.1 Also, $g_n(x)$ converges to $g(x)$ for each $x \in X$. From (3.3.3) it follows that

$$|g_n(V)g_n(Z)| \le ||g||^2\, m^2\, ||V||\ ||Z|| \qquad (3.3.4)$$

uniformly in n and $\omega \in \Omega$. Moreover,

$$E(||g||^2 m^2 ||V||\ ||Z||) \le (m||g||)^2 (E||V||^2)^{1/2} (E||Z||^2)^{1/2} < \infty.$$

Thus, by the dominated convergence theorem

$$E[g(V)g(Z)] = E[\lim_{n\to\infty} g_n(V)g_n(Z)] = \lim_{n\to\infty} E[g_n(V)g_n(Z)]. \qquad (3.3.5)$$

But,

$$g_n(V) = g(U_n(V)) = g(\textstyle\sum_{k=1}^n f_k(V)b_k) = \sum_{k=1}^n f_k(V)g(b_k),$$

and

$$g_n(Z) = \textstyle\sum_{k=1}^n f_k(Z)g(b_k).$$

Hence, for each n

$$E[g_n(V)g_n(Z)] = \sum_{k=1}^{n-1} \sum_{\ell=k+1}^n g(b_k)g(b_\ell)E[f_k(V)f_\ell(Z) + f_\ell(V)f_k(Z)]$$
$$+ \sum_{k=1}^n [g(b_k)]^2\, E[f_k(V)f_k(Z)]. \qquad (3.3.6)$$

Condition (3.3.1) implies that each term of each sum is zero and hence that $E[g_n(V)g_n(Z)] = 0$ for each n. Therefore, (3.3.2) is satisfied since

$$E[g(V)g(Z)] = \lim_{n\to\infty} E[g_n(V)g_n(Z)] = 0,$$

and the random elements V and Z are weakly uncorrelated. ///

Theorem 3.3.1 is also valid for any normed linear space which has a Schauder basis such that $\{||U_n||\}$ is bounded. The verification of Condition (3.3.1) depends on the particular basis $\{b_n\}$ which is used, but it must hold for every basis for the random elements to be weakly uncorrelated. Also observe that Condition (3.3.1) must hold for k = n, or for each k

$$\mathrm{Cov}\,[f_k(V),\ f_k(Z)] = 0.$$

Thus, weak uncorrelation implies coordinate uncorrelation for every Schauder basis where the coordinate functionals are continuous.

For the random elements $V = (V_1, V_2, \ldots)$ and $Z = (Z_1, Z_2, \ldots)$ [with $E||V||^2 < \infty$ and $E||Z||^2 < \infty$] in the sequence spaces $c, c_0, R^{(\infty)}$, and $\ell^p (p \geq 1)$, Theorem 3.3.1 states that V and Z are weakly uncorrelated if and only if

$$\mathrm{Cov}\,(V_n + V_k, Z_n + Z_k) = 0 \tag{3.3.7}$$

for each k and n. Condition (3.3.7) is also necessary and sufficient for random elements to be weakly uncorrelated in c. This is easily verified for c_0, $R^{(\infty)}$, or $\ell^p (p \geq 1)$ by considering the basis $\{(1,0,0,\ldots), (0,1,0,\ldots), \ldots\}$. For the space c, recall that $\{(1,1,1,\ldots), (1,0,0,\ldots), (0,1,0,\ldots), \ldots\}$ is a basis and the coordinate functionals $\{f_n\}$ are given by $f_0(x) = \lim x_m$ and $f_n(x) = x_n - \lim x_m$ for each $n \geq 1$. To show that Condition (3.3.7) implies Condition (3.3.1), three cases are considered. First, without loss of generality, assume that $EV = 0 = EZ$. Thus, $E(\lim V_m) = 0 = E(\lim Z_m)$ and $E(V_n) = 0 = E(Z_n)$ for each n.

Case 1: $k = 0$ and $n = 0$

First,

$$\mathrm{Cov}[f_0(V) + f_0(V), f_0(Z) + f_0(Z)] = 4E[(\lim V_m)(\lim Z_m)].$$

But, for each m

$$|V_m Z_m| \leq ||V||\ ||Z||,$$

and

$$E(||V||\ ||Z||) \leq (E||V||^2)^{\frac{1}{2}}(E||Z||^2)^{\frac{1}{2}} < \infty. \tag{3.3.8}$$

Hence,

$$E[(\lim V_m)(\lim Z_m)] = E[\lim (Z_m V_m)]$$

$$= \lim E(Z_m V_m) = 0 \qquad (3.3.9)$$

since Condition (3.3.7) with m = n = k implies

$$0 = \text{Cov}\,(V_n + V_k, Z_n + Z_k) = 4E\,(V_m Z_m).$$

Case 2: k = 0 and n ≠ 0 (similar for k ≠ 0 and n = 0).

Note that

$$\text{Cov}\,[f_n(V) + f_0(V),\ f_n(Z) + f_0(Z)]$$

$$= \text{Cov}\,[V_n - \lim V_m + \lim V_m,\ Z_n - \lim Z_m + \lim Z_m]$$

$$= E[V_n Z_n] = 0$$

by Condition (3.3.7).

Case 3: n ≠ 0 and k ≠ 0

Now,

$$\text{Cov}\,[f_n(V) + f_k(V),\ f_n(Z) + f_k(Z)]$$

$$= \text{Cov}\,[V_n - \lim V_m + V_k - \lim V_m,\ Z_n - \lim Z_m + Z_k - \lim Z_m]$$

$$= E[((V_n + V_k) - 2 \lim V_m)((Z_n + Z_k) - 2 \lim Z_m)]$$

$$= E[(V_n + V_k)(Z_n + Z_k)] + 4E[(\lim V_m)(\lim Z_m)]$$

$$- 2[E(V_n \lim Z_m) + E(V_k \lim Z_m) + E(Z_n \lim V_m) + E(Z_k \lim V_m)]. \qquad (3.3.10)$$

Condition (3.3.7) implies that $E[(V_n + V_k)(Z_n + Z_k)] = 0$ while Case 1 implies that $E[(\lim V_m)(\lim Z_m)] = 0$. For each n

$$0 = \lim_{m\to\infty} E[V_n Z_m + Z_n V_m] = E[V_n \lim Z_m + Z_n \lim V_m]$$

using $|V_n Z_m + Z_n V_m| \le 2||V||\ ||Z||$ and Inequality (3.3.8).

Similarly,

$$0 = E[V_k \lim Z_m] + E[Z_k \lim V_m].$$

Thus, in any case Cov $[f_n(V) + f_k(Z), f_n(Z) + f_k(Z)] = 0.$ ///

For the space C[0,1] a little work is required to describe weakly uncorrelated random elements since the dual space is isometrically isomorphic to the space of all finite signed Baire measures on [0,1], Royden (1972), page 257. Recall that V is a random element in C[0,1] if and only if V(t) is a random variable for each $t \in [0,1]$ and V(t) is a continuous function of t for each $\omega \in \Omega$. Using Fubini's Theorem it can be proved directly that the random elements V and Z (with $E||V||^2 < \infty$ and $E||Z||^2 < \infty$) are weakly uncorrelated if and only if for all $s,t \in [0,1]$

$$\text{Cov}[V(t) + V(s), Z(t) + Z(s)] = 0. \qquad (3.3.11)$$

On the other hand, Theorem 3.3.1 can be used to characterize weakly uncorrelated random elements in C[0,1] when a basis is given. A basis and a set of coordinate functionals for C[0,1] are given by Example 1.3.1. Condition (3.3.11) with the appropriate modifications is also necessary and sufficient for weakly uncorrelation in $C[0,\infty)$.

It is clear that the usefulness of these characterizations depends on the particular basis which is used. The next result will show that random elements which are coordinate uncorrelated with respect to an orthonormal basis of a Hilbert space are uncorrelated. Since weakly uncorrelated random elements must be coordinate uncorrelated for every Schauder basis in a Banach space, weakly uncorrelated random elements must be uncorrelated in a separable Hilbert space when the appropriate moment conditions hold.

Theorem 3.3.2 Let H be a separable Hilbert space and let V and Z be random elements in H with $E||V||^2 < \infty$ and $E||Z||^2 < \infty$. If V and Z are coordinate uncorrelated with respect to an orthonormal basis $\{b_n\}$, then V and Z are uncorrelated.

Proof: Again, it can be assumed that EV = 0 = EZ. From Definitions 2.3.6 and 3.3.2, it follows that

$$E(\langle V, b_n\rangle\langle Z, b_n\rangle) = 0 \tag{3.3.12}$$

for each n. By Remark 1.3.1 the random variable $\langle V,Z\rangle$ can be expressed as

$$\langle V,Z\rangle = \sum_{n=1}^{\infty} \langle V, b_n\rangle\langle Z, b_n\rangle. \tag{3.3.13}$$

Similarly to (3.2.6), it follows that for each m

$$\sum_{n=1}^{m} |\langle V, b_n\rangle\langle Z, b_n\rangle| \le ||V||\ ||Z||$$

and that

$$E(||V||\ ||Z||) \le (E||V||^2)^{\frac{1}{2}}(E||Z||^2)^{\frac{1}{2}} < \infty.$$

Again by the dominated convergence theorem,

$$E\langle V,Z\rangle = E(\lim_{m\to\infty} \sum_{n=1}^{m} \langle V, b_n\rangle\langle Z, b_n\rangle)$$
$$= \lim_{m\to\infty} \sum_{n=1}^{m} E(\langle V, b_n\rangle\langle Z, b_n\rangle) = 0.$$

Thus, V and Z are uncorrelated random elements. ///

The following example will show that no implications exists between coordinate uncorrelated random elements and uncorrelated random elements. In particular, it will show that the converse of Theorem 3.3.2 need not hold and that random elements may be coordinate uncorrelated but not be uncorrelated.

<u>Example 3.3.1</u> Let $X = R^2$ with the usual inner product and let A_1 and A_2 be independent random variables such that $EA_1 = 0 = EA_2$ and $E(A_1^2) = 1 = E(A_2^2)$. The random elements $V = (A_1, A_2)$ and $Z = (A_1, -A_2)$ are obviously not coordinate uncorrelated with respect to the orthonormal basis $\{(1,0), (0,1)\}$, but

$$E\langle V,Z\rangle = E(A_1^2 - A_2^2) = 0 = \langle 0,0\rangle = \langle EV, EZ\rangle.$$

Thus, uncorrelated random elements need not be coordinate uncorrelated for each orthonormal basis for X.

Next, consider the random elements $V = (A_1, A_2)$ and $Z = (A_2, A_2)$. Obviously, the random elements V and Z are not uncorrelated since

$$\begin{aligned} E\langle V,Z\rangle &= E(A_1A_2 + A_2A_2) \\ &= (EA_1)(EA_2) + E(A_2^2) = 1 \neq 0 = \langle EV, EZ\rangle. \end{aligned}$$

But, for the Schauder basis $\{(1,1), (0,1)\}$, the random elements are coordinate uncorrelated since

$$E[f_1(V)f_1(Z)] = E[A_1A_2] = 0$$

and

$$E[f_2(V)f_2(Z)] = E[(A_2 - A_1)(A_2 - A_2)] = E[0] = 0.$$

Thus, random elements can be coordinate uncorrelated without being uncorrelated. ///

In the definition of uncorrelated random elements (Definition 3.2.1), it was assumed that $E||V||^2 < \infty$ and $E||Z||^2 < \infty$. However, in the definitions of coordinate uncorrelated (Definition 3.3.2) and weakly uncorrelated random elements (Definition 3.3.1), it was only assumed that $E[f_k(V)^2] < \infty$ and $E[f_k(Z)^2] < \infty$ and that $E[f(V)^2] < \infty$ and $E[f(V)^2] < \infty$. It is not difficult to construct weakly uncorrelated random elements in a separable Hilbert space whose second absolute moments are not finite. For example, let H be the separable Hilbert space ℓ^2 and define V to be $n\delta^n$ with probability c/n^2 where c is the appropriate constant and $\delta^n = (0,\ldots,0,1,0,\ldots)$ with the "1" occurring in the nth coordinate. Note,

$$E||V||^2 = \sum_{n=1}^{\infty} n^2 c/n^2 = \infty,$$

but for each $f = (t_1,t_2,\ldots) \in (\ell^2)^* = \ell^2$

$$E[f(V)^2] = \sum_{n=1}^{\infty} n^2 t_n^2 \; c/n^2 = c \sum_{n=1}^{\infty} t_n^2 < \infty.$$

Then let Z be a random element in ℓ^2 with the same probability law as V but which is independent of V.

3.4 CONVERGENCE OF WEIGHTED SUMS OF RANDOM VARIABLES

A review of some of the existing results on convergence of weighted sums of random variables will be presented in this section. These results will be used primarily as reference in the later presentation of results for linear topological spaces. Thus, a rather exhaustive listing will be attempted, but proofs will not be included. The reader can find the proofs and additional details in the various journals where the results have recently appeared.

Definition 3.4.1 A double array $\{a_{nk}: n,k = 1,2,\ldots\}$ of real numbers is said to be a Toeplitz sequence, T sequence, if

$$\text{(i)} \quad \lim_{n\to\infty} a_{nk} = 0 \text{ for each } k, \tag{3.4.1}$$

and

$$\text{(ii)} \quad \sum_{k=1}^{\infty} |a_{nk}| \le C \text{ for each } n. \tag{3.4.2}$$

The constant C is often assumed to be 1 (without loss of generality) in the stochastic convergence results. The following deterministic lemma on Toeplitz sequences is from Loéve (1963), page 238.

Lemma 3.4.1 Let $\{x_n\}$be a sequence of real numbers and let $\{a_{nk}\}$ be a Toeplitz sequence.

(a) If $x_n \to 0$, then $\sum_{k=1}^{n} a_{nk}x_k \to 0$.

(b) If $x_n \to x \in R$ and $\sum_{k=1}^{n} a_{nk} \to 1$, then $\sum_{k=1}^{n} a_{nk}x_k \to x$.

Finally, if $\{a_k\} \subset R^+$ such that $b_n = \sum_{k=1}^{n} a_k \to \infty$, then

$$\frac{1}{b_n} \sum_{k=1}^{n} a_k x_k \to x.$$

The Toeplitz sequence

$$a_{nk} = \begin{cases} \frac{1}{n} & \text{if } k = 1,\ldots,n \\ 0 & \text{if } k > n \end{cases}$$

is often used. In this case, Lemma 3.4.1 yields that convergence implies Cesaro convergence. A choice of $a_k = 1$ for all k yields the same implication since $b_n = n$.

Jamison, Orey, and Pruitt (1965) considered independent, identically distributed random variables and weights $\{a_k\} \subset R^+$. Since

$$\frac{1}{b_n} \sum_{k=1}^{n} a_k X_k = \left(\frac{b_{n-1}}{b_n}\right) \frac{1}{b_{n-1}} \sum_{k=1}^{n-1} a_k X_k + \frac{a_n}{b_n} X_n,$$

convergence in probability to a constant for non-degenerate random variables necessitates that $\lim_{n\to\infty} \frac{a_n}{b_n} = 0$. Thus, assuming that

$$\max_{1 \le k \le n} \left(\frac{a_k}{b_n}\right) \to 0 \text{ and } b_n \to \infty, \tag{3.4.3}$$

the following three results were obtained by Jamison, Orey, and Pruitt (1965).

Theorem 3.4.2 For independent, identically distributed random variables $\{X_k\}$

$$\lim_{t\to\infty} t\, P[|X_1| \geq t] = 0 \text{ and } \lim_{t\to\infty} \int_{[|X_1|<t]} X_1 dP \text{ exists}$$

if and only if

$$\frac{1}{b_n} \sum_{k=1}^{n} a_k X_k \to c$$

in probability for some constant c.

Theorem 3.4.3 Let $\{X_k\}$ be independent, identically distributed random variables and let N(y) be the number of n such that $b_n/a_n \leq y$. If $E|X_1| < \infty$ and

$$\int x^2 \int_{y\geq|x|} \frac{N(y)}{y^3}\, dy dF_{X_1}(x) < \infty, \tag{3.4.4}$$

then

$$\frac{1}{b_n} \sum_{k=1}^{n} a_k X_k \to c$$

with probability one for some constant c.

Condition (3.4.4) implies that $E[N(|X_1|)] < \infty$.

Theorem 3.4.4 If $\{X_k\}$ are independent, identically distributed random variables such that $E(|X_1|\log^+|X_1|) < \infty$, then

$$\frac{1}{b_n} \sum_{k=1}^{n} a_k X_k \to c$$

with probability one for some constant c.

The following two theorems are from Pruitt (1966) and are for independent, identically distributed random variables and Toeplitz sequences.

Theorem 3.4.5 Let $\{X_k\}$ be independent, identically distributed random variables such that $E|X_1| < \infty$ and let $\{a_{nk}\}$ be a Toeplitz sequence such that $\lim_{n\to\infty} \sum_{k=1}^{n} a_{nk} = 1$. A necessary and sufficient condition that

$$Y_n = \sum_{k=1}^{\infty} a_{nk}X_k \to EX_1 \quad (3.4.5)$$

in probability is that

$$\max_k |a_{nk}| \to 0. \quad (3.4.6)$$

A sequence $\{t_n\}$ is said to be $\mathcal{O}(f(n))$ if $\sup_n |\frac{t_n}{f(n)}| < \infty$ and is said to be $o(f(n))$ if $\lim_{n\to\infty} |\frac{t_n}{f(n)}| = 0$.

Theorem 3.4.6 Let $\{X_k\}$ be independent, identically distributed random variables and let $\{a_{nk}\}$ be a Toeplitz sequence. If $\max_k |a_{nk}| = O(n^{-\gamma})$ for some $\gamma > 0$, then $E|X_1|^{1+1/\gamma} < \infty$ implies that

$$Y_n = \sum_{k=1}^{\infty} a_{nk}X_k \to EX_1$$

with probability one.

Let $\{a_{nk}\}$ be an array of real numbers such that

$$A_n = \sum_{k=1}^{\infty} a_{nk}^2 < \infty \quad (3.4.7)$$

for each n. Chow (1966) defined a generalized Gaussian random variable and showed that bounded random variables with zero means were generalized Gaussian. Obviously, Gaussian random variables are generalized Gaussian.

Definition 3.4.2 [Chow (1966)] A random variable X is said to be generalized Gaussian if there exists $\alpha \geq 0$ such that

$$E[\exp(tX)] \leq \exp(\alpha^2 t^2/2) \quad (3.4.8)$$

for all $t \in R$. When (3.4.8) holds, let

$$\tau(X) = \inf\{\alpha \geq 0: \ (3.4.8) \text{ holds}\}.$$

The following two theorems are from Chow (1966).

Theorem 3.4.7 Let $\{X_{nk}: k = 1,2,\ldots\}$ be independent random variables for each n and let $X_{nk} - EX_{nk}$ be generalized Gaussian with $\sup_k \tau(X_{nk} - EX_{nk}) \le \sqrt{2}$ for each n. Let

$$C = \lim_{n\to\infty} C_n = \lim_{n\to\infty} \sum_{k=1}^{\infty} a_{nk}EX_{nk} \tag{3.4.9}$$

exist. If for every $\varepsilon > 0$

$$\sum_{n=1}^{\infty} \exp(-\varepsilon/A_n) < \infty \tag{3.4.10}$$

or if $A_n = o(\log^{-1} n)$, then

$$Y_n = \sum_{k=1}^{\infty} a_{nk}X_{nk} \to C$$

with probability one.

Theorem 3.4.8 Let $\{X_n\}$ be independent, identically distributed random variables. Then $EX_1 = 0$ and $E(X_1^2) < \infty$ if and only if for every array $\{a_{nk}\}$ such that $\lim_{n\to\infty} \sum_{k=1}^{n} a_{nk}^2 = 1$ it follows that

$$\frac{1}{\sqrt{n}} \sum_{k=1}^{n} a_{nk}X_k \to 0$$

with probability one.

Stout (1968) considered the complete convergence of the sequence $\{Y_n\}$ in the sense of

$$\sum_{n=1}^{\infty} P[|Y_n| > \varepsilon] < \infty \tag{3.4.11}$$

for each $\varepsilon > 0$. Note that Condition (3.4.11) implies that $Y_n \to 0$ with probability one. Stout's (1968) results involved varying moment conditions and conditions on the array $\{a_{nk}\}$ in addition to (3.4.9) and (3.4.10) of Chow (1966).

Rohatgi (1971) extended Pruitt's (1966) results to independent, but not necessarily identically distributed random variables $\{X_k\}$ by requiring that there exists a random variable X such that

$$P[|X_n| \geq t] \leq P[|X| \geq t] \qquad (3.4.12)$$

for all $t > 0$ and all n. Theorems 3.4.9 and Theorem 3.4.10 are from Rohatgi (1971).

<u>Theorem 3.4.9</u> Let $\{X_n\}$ be a sequence of independent random variables such that Condition (3.4.12) holds and let $\{a_{nk}\}$ be a Toeplitz sequence. If

(i) $\max_k |a_{nk}| \to 0$ as $n \to \infty$,

(ii) $E|X|^r < \infty$ for some $0 < r \leq 1$, and

(iii) $\sum_{k=1}^{\infty} |a_{nk}|^r \leq C$ for each n,

then

$$Y_n = \sum_{k=1}^{\infty} a_{nk}X_k \to 0$$

in probability (if $r = 1$, then also assume that $EX_n = 0$ for all n).

<u>Theorem 3.4.10</u> Let $\{X_n\}$ be a sequence of independent random variables such that $EX_n = 0$ for all n and such that Condition (3.4.12) holds and let $\{a_{nk}\}$ be a Toeplitz sequence. If

(i) $\max_k |a_{nk}| = \mathcal{O}(n^{-\gamma})$ for some $\gamma > 0$

and

(ii) $E|X|^{1+1/\gamma} < \infty$,

then

$$Y_n = \sum_{k=1}^{\infty} a_{nk}X_k \to 0$$

with probability one.

Chow and Lai (1973) considered the convergence of

$$Y_n = \sum_{i=1}^{n} c_{n-i}X_i$$

where $\{X_n\}$ are independent, identically distributed random variables

$$\sum_{n=0}^{\infty} c_n^{\,2} < \infty.$$

Such weighted sums of observations are used in Lai (1972) for detecting changes in the location of the distribution of a sequence of independent observations, such as in quality control problems. Theorems 3.4.11, 3.4.12, and 3.4.13 are from Chow and Lai (1973).

<u>Theorem 3.4.11</u> Let $\{X_n\}$ be a sequence of independent, identically distributed random variables such that $EX_1 = 0$. For $\alpha \geq 1$, the following statements are equivalent:

(i) $E|X_1|^\alpha < \infty$;

(ii) $\lim_{n\to\infty} n^{-1/\alpha} X_n = 0$ with probability one;

(iii) $\lim_{n\to\infty} n^{-1/\alpha} \sum_{i=1}^{n} c_{n-i} X_i = 0$ with probability one.

<u>Theorem 3.4.12</u> Let $\{X_n\}$ be independent, identically distributed random variables such that $EX_1 = 0$. Then the following statements are equivalent:

(i) $E[\exp(t|X_1|)] < \infty$ for all $t > 0$;

(ii) $\lim_{n\to\infty} X_n/\log n = 0$ with probability one;

(iii) $\lim_{n\to\infty} \sum_{i=1}^{n} c_{n-i} X_i/\log n = 0$ with probability one.

<u>Theorem 3.4.13</u> Let $\{X_n\}$ be independent, identically distributed random variables such that $EX_1 = 0$. For $1 \leq \alpha \leq 2$, $E|X_1|^\alpha < \infty$ if and only if for every array $\{a_{nk}\}$ such that $\limsup_{n\to\infty} \sum_{k=1}^{n} a_{nk}^2 < \infty$, it follows that

$$n^{-1/\alpha} \sum_{k=1}^{n} a_{nk} X_k \to 0$$

with probability one.

Wright, Platt, and Robertson (1977) continued the investigation of Jamison, Orey and Pruitt (1965) on the convergence of

$$\frac{1}{b_n} \sum_{k=1}^{n} a_k X_k$$

where $b_n = a_1 + \ldots + a_n$ and $\{X_n\}$ are independent, identically distributed random variables. In particular, Wright, Platt, and Robertson (1977) showed that in Theorem 3.4.3 the condition that $E|X_1| < \infty$ could be replaced by

$$\int_{|x|<T} x dF_{X_1}(x) \to c \text{ as } T \to \infty.$$

3.5 PROBLEMS

3.1 Let X be a normed linear space and let V be a random element in X. Show that $E(||V||^2) < \infty$ implies that $E||V|| < \infty$.

3.2 In a Hilbert space H verify that

$$\langle x,y\rangle = \tfrac{1}{4}(||x + y||^2 - ||x - y||^2), \text{ for all } x,y \in H.$$

3.3 Let V and Z be random elements in C[0,1] such that $E||V||^2 < \infty$ and $E||Z||^2 < \infty$. Show that V and Z are weakly uncorrelated if and only if

$$\text{Cov}\,[V(t) + V(s), Z(t) + Z(s)] = 0$$

for all $s,t \in [0,1]$ including $s = t$. [C[0,1]* is the finite signed Baire measures on [0,1], that is, $\mu \in C[0,1]^*$ implies $\mu(x) = \int_0^1 x(t)\, d\mu(t)$ where μ is a finite signed Baire measure.]

3.4 Show that random elements V and Z are weakly uncorrelated in s if and only if

$$\text{Cov}\,(V_n + V_k, Z_n + Z_k)$$

for all n and k. [s* is $R^{(\infty)}$.]

3.5 In Example 3.3.1, are the random elements $V = (A_1,A_2)$ and $Z = (A_1,-A_2)$ coordinate uncorrelated with respect to the basis $\{(1,1), (1,-1)\}$?

CHAPTER IV

LAWS OF LARGE NUMBERS IN NORMED LINEAR SPACES

4.0 INTRODUCTION

The most classical form of a weighted sum is the mean (or average) where the weights from 1 to n are uniformly 1/n. The stochastic convergence of the mean, or the laws of large numbers, in Banach spaces has been studied extensively in the last two decades. The laws of large numbers for identically distributed random variables were extended to normed linear spaces by Mourier (1953) and Taylor (1972). However, additional restrictions on the distributions or on the Banach spaces are needed to obtain the corresponding results for the non-identically distributed random elements.

In this chapter several of the laws of large numbers for random elements in normed linear spaces will be presented. While page limitations will not permit an exhaustive presentation of these results, sufficient information is presented to provide an overview of the area and references to results will indicate where additional details are available. Proofs of several of these results will be included to illustrate some of the techniques and problems in generalizing laws of large numbers to normed linear spaces. In addition, examples are given which show that many of the seemingly obvious extensions are not valid for normed linear spaces.

4.1 LAWS OF LARGE NUMBERS FOR IDENTICALLY DISTRIBUTED RANDOM ELEMENTS

In this section laws of large numbers will be obtained for identically distributed random elements in separable normed linear spaces. The first result to be presented in this section will be Mourier's (1953) strong law of large numbers for a sequence of independent, identically distributed random elements $\{V_n\}$ in a separable Banach space where $E||V_1|| < \infty$.

Recall from Proposition 1.2.1 that for a separable normed linear space there exists a countably-valued Borel measurable function T_m from X into X such that $||T_m x - x|| < 1/m$ for each $m = 1,2,\ldots$ and each $x \in X$.

Theorem 4.1.1 If X is a separable Banach space and $\{V_n\}$ is a sequence of independent, identically distributed random elements in X such that $E||V_1|| < \infty$, then

$$||\frac{1}{n} \sum_{k=1}^{n} V_k - EV_1|| \to 0$$

with probability one.

Proof: Since $E||V_1|| < \infty$ and X is separable and complete, EV_1 exists by Lemma 2.3.8. Moreover, $EV_n = EV_1$ for each n since the random elements $\{V_n\}$ are identically distributed.

First, assume that the random elements $\{V_n\}$ can take only countably many values $x_1, x_2, \ldots$. For each positive integer t define

$$V_k^t = \sum_{i=1}^{t} x_i I_{[V_k = x_i]} \tag{4.1.1}$$

and define $R_k^t = V_k - V_k^t$ for each k. For each i the random variables $\{I_{[V_k=x_i]}: k \geq 1\}$ are independent and identically distributed with $E[I_{[V_k=x_i]}] = P[V_k = x_i]$. Moreover, for each k

$$EV_k^t = \sum_{i=1}^t x_i P[V_1 = x_i] = EV_1^t$$

by Theorem 2.3.7 (i) and (iii). Hence, a finite-dimensional version of Theorem 3.1.5 implies that

$$\begin{aligned} 0 &\le \left\|\frac{1}{n}\sum_{k=1}^n V_k^t - EV_1^t\right\| \\ &= \left\|\frac{1}{n}\sum_{k=1}^n \sum_{i=1}^t x_i I_{[V_k=x_i]} - \sum_{i=1}^t x_i P[V_1 = x_i]\right\| \\ &\le \sum_{i=1}^t \|x_i\| \left|\frac{1}{n}\sum_{k=1}^n I_{[V_k=x_i]} - P[V_1 = x_i]\right| \to 0 \qquad (4.1.2) \end{aligned}$$

with probability one for each t. For each t, $\{\|R_n^t\|\}$ is a sequence of independent, identically distributed random variables such that $E\|R_1^t\| \le E\|V_1\| < \infty$. Thus,

$$\frac{1}{n}\sum_{k-1}^n \|R_k^t\| \to E\|R_1^t\| \qquad (4.1.3)$$

with probability one for each t. Since $\|R_1^t\| \to 0$ pointwise as $t \to \infty$ and $\|R_1^t\| \le \|V_1\|$ for each t, the Lebesgue dominated convergence theorem implies that $E\|R_1^t\| \to 0$ as $t \to \infty$.

Let S be the countable union of null sets for which (4.1.2) and (4.1.3) do not hold. Let $\omega \in S$ and let $\varepsilon > 0$ be given. Choose t such that

$$E\|R_1^t\| \le \frac{\varepsilon}{4}. \qquad (4.1.4)$$

By (4.1.2) and (4.1.3) there exists a positive integer $N(\varepsilon,\omega)$ such that for all $n \ge N(\varepsilon,\omega)$

$$\left\|\frac{1}{n}\sum_{k=1}^n V_k^t(\omega) - EV_1^t\right\| < \frac{\varepsilon}{4} \qquad (4.1.5)$$

and

$$\frac{1}{n}\sum_{k=1}^n \|R_k^t(\omega)\| < E\|R_1^t\| + \frac{\varepsilon}{4} \le \frac{\varepsilon}{2} \qquad (4.1.6)$$

from (4.1.4). By Theorem 2.3.7 (i) and (v)

$$||EV_1^t - EV_1|| \le E||V_1^t - V_1|| = E||R_1^t|| \le \frac{\varepsilon}{4}. \qquad (4.1.7)$$

Hence, for $\omega \notin S$ and $n \ge N(\varepsilon,\omega)$

$$\begin{aligned} ||\tfrac{1}{n}\textstyle\sum_{k=1}^{n} V_k(\omega) - EV_1|| &\le ||\tfrac{1}{n}\textstyle\sum_{k=1}^{n} V_k^t(\omega) - EV_1^t|| \\ &\quad + \tfrac{1}{n}\textstyle\sum_{k=1}^{n} ||R_k^t(\omega)|| + ||EV_1^t - EV_1|| \\ &< \tfrac{\varepsilon}{4} + \tfrac{\varepsilon}{2} + \tfrac{\varepsilon}{4} = \varepsilon. \end{aligned}$$

Hence, the theorem is proved for countably-valued random elements.

In the general case

$$\begin{aligned} ||\tfrac{1}{n}\textstyle\sum_{k=1}^{n} V_k - EV_1|| &\le \tfrac{1}{n}\textstyle\sum_{k=1}^{n}||V_k - T_mV_k|| \\ &\quad + ||\tfrac{1}{n}\textstyle\sum_{k=1}^{n} T_mV_k - ET_mV_1|| + ||ET_mV_1 - EV_1|| \end{aligned} \qquad (4.1.8)$$

for each n and m. By the first part of the proof

$$||\tfrac{1}{n}\textstyle\sum_{k=1}^{n} T_mV_k - ET_mV_1|| \to 0 \qquad (4.1.9)$$

with probability one for each m since the random elements $\{T_mV_n:\ n \ge 1\}$ are independent, identically distributed, countably-valued, and $E||T_mV_1|| \le E||V_1|| + \frac{1}{m} < \infty$. Moreover, for each n

$$\tfrac{1}{n}\textstyle\sum_{k=1}^{n} ||T_mV_k - V_k|| \le \tfrac{1}{m}$$

and

$$||ET_mV_1 - EV_1|| \le E||T_mV_1 - V_1|| \le \tfrac{1}{m}$$

since $||T_mx - x|| \le \frac{1}{m}$ for each $x \in X$. The proof is completed by letting the null set be the countable union of null sets for which the convergence in (4.1.9) does not hold. ///

Each normed linear space is isomorphic to a dense subset of a Banach space [Horvath (1966), page 25]. Hence, Theorem 4.1.1 can easily be extended to separable normed linear spaces by assuming the existence of EV_1 and by using Theorem 2.3.7.

Corollary 4.1.2 If X is a separable normed linear space and $\{V_n\}$ is a sequence of independent, identically distributed random elements in X such that EV_1 exists and $E||V_1|| < \infty$, then

$$||\frac{1}{n} \sum_{k=1}^{n} V_k - EV_1|| \to 0$$

with probability one.

In addition to the problem of defining an appropriate concept of uncorrelation, the methods of proof for weak laws of large numbers must be different since Borel measurable functions (or truncation to a countably number of values) of uncorrelated (in some mode) random elements may not be uncorrelated. However, Taylor (1972) showed that a sequence of identically distributed random elements in a Banach space which has a Schauder basis satisfies the weak law of large numbers in the norm topology if and only if the weak law of large numbers holds in each coordinate of the basis. Hence, weak laws of large numbers for coordinate uncorrelated random elements can be obtained as corollaries from the following theorem.

Theorem 4.1.3 Let X be a Banach space which has a Schauder basis $\{b_n\}$ and let $\{V_n\}$ be a sequence of identically distributed random elements in X such that $E||V_1|| < \infty$. For each coordinate functional f_i, the weak law of large numbers holds for the random variables $\{f_i(V_n): n \geq 1\}$ if and only if

$$||\frac{1}{n} \sum_{k=1}^{n} V_k - EV_1|| \to 0$$

in probability.

Proof: The "if" part is obvious since convergence in the norm topology implies convergence in the weak linear topology of X and in each coordinate. The "only if" part is proved below.

Since $E||V_1|| < \infty$ and X is complete and separable, EV_1 exists and can assumed to be 0 (otherwise, consider $Z_n = V_n - EV_1$). Let $\varepsilon > 0$ and $\delta > 0$ be given. To show that

$$||\frac{1}{n} \sum_{k=1}^{n} V_k|| \to 0$$

in probability, there must exist a positive integer $N(\varepsilon,\delta)$ such that

$$P[||\frac{1}{n} \sum_{k=1}^{n} V_k|| > \varepsilon] < \delta \tag{4.1.10}$$

for each $n \geq N(\varepsilon,\delta)$.

Let $m > 0$ be the basis constant given in Lemma 1.3.1 such that $||U_t|| \leq m$ for all t. Hence, $||Q_t|| \leq m + 1$ for all t where $Q_t(x) = x - U_t(x)$. For each n and each t

$$\frac{1}{n} \sum_{k=1}^{n} V_k = \frac{1}{n} \sum_{k=1}^{n} U_t(V_k) + \frac{1}{n} \sum_{k=1}^{n} Q_t(V_k). \tag{4.1.11}$$

Also for each t and each n

$$\begin{aligned} P[||\frac{1}{n} \sum_{k=1}^{n} Q_t(V_k)|| > \frac{\varepsilon}{2}] &\leq P[\frac{1}{n} \sum_{k=1}^{n} ||Q_t(V_k)|| > \frac{\varepsilon}{2}] \\ &\leq \frac{2}{\varepsilon n} \sum_{k=1}^{n} E||Q_t(V_k)|| \\ &= \frac{2}{\varepsilon} E||Q_t(V_1)||. \end{aligned} \tag{4.1.12}$$

But $E||Q_t(V_1)|| \to 0$ as $t \to \infty$ since $||Q_t(V_1)|| \to 0$ pointwise and $||Q_t(V_1)|| \leq (m + 1)||V_1||$. Thus, choose t so that

$$P[||\frac{1}{n} \sum_{k=1}^{n} Q_t(V_k)|| > \frac{\varepsilon}{2}] < \frac{\delta}{2} . \tag{4.1.13}$$

Next,

$$\begin{aligned} P[||\frac{1}{n} \sum_{k=1}^{n} U_t(V_k)|| > \frac{\varepsilon}{2}] &= P[||\sum_{i=1}^{t} f_i(\frac{1}{n} \sum_{k=1}^{n} V_k) b_i|| > \frac{\varepsilon}{2}] \\ &\leq P[\sum_{i=1}^{t} |f_i(\frac{1}{n} \sum_{k=1}^{n} V_k)| \; ||b_i|| > \frac{\varepsilon}{2}] \\ &\leq \sum_{i=1}^{t} P[|\frac{1}{n} \sum_{k=1}^{n} f_i(V_k)| > \frac{\varepsilon}{2t||b_i||}] \end{aligned} \tag{4.1.14}$$

where $\{f_i\}$ are the coordinate functionals and $\{b_i\}$ is the Schauder basis. But, for each i

$$P[|\frac{1}{n}\sum_{k=1}^{n} f_i(V_k)| > \frac{\varepsilon}{2t||b_i||}] \to 0$$

as $n \to \infty$ since the weak law of large numbers holds for each sequence $\{f_i(V_n): n \geq 1\}$ and since $E[f_i(V_1)] = 0$. Hence, there exists a positive integer $N(\varepsilon,\delta)$ such that

$$P[||\frac{1}{n}\sum_{k=1}^{n} U_t(V_k)|| > \frac{\varepsilon}{2}] < \frac{\delta}{2} \tag{4.1.15}$$

for each $n \geq N(\varepsilon,\delta)$. Using (4.1.11), (4.1.13), and (4.1.15), it follows that

$$P[||\frac{1}{n}\sum_{k=1}^{n} V_k|| > \varepsilon]$$

$$\leq P[||\frac{1}{n}\sum_{k=1}^{n} U_t(V_k)|| > \frac{\varepsilon}{2}] + P[||\frac{1}{n}\sum_{k=1}^{n} Q_t(V_k)|| > \frac{\varepsilon}{2}]$$

$$< \frac{\delta}{2} + \frac{\delta}{2} = \delta$$

for each $n \geq N(\varepsilon,\delta)$. Thus, (4.1.10) is obtained, and hence

$$||\frac{1}{n}\sum_{k=1}^{n} V_k|| \to 0$$

in probability. ///

The same proof also yields the result for any separable normed linear space which has a Schauder basis such that $\{||U_n||\}$ is a bounded sequence, but the existence of EV_1 must be assumed for an incomplete space. Thus, Theorem 4.1.3 is valid for $R^{(\infty)}$. It is important to observe that Theorem 4.1.3 states that for identically distributed random elements the weak law of large numbers holds in the norm topology of a Banach space if there exists some Schauder basis such that the weak law of large numbers holds in each coordinate. Thus, if there exists a Schauder basis such that the identically distributed random elements are coordinate uncorrelated, then the weak law of large numbers holds.

Corollary 4.1.4 If X is a Banach space and if $\{V_n\}$ is a sequence of identically distributed, coordinate uncorrelated random elements in X such that $E||V_1|| < \infty$, then

$$||\frac{1}{n} \sum_{k=1}^{n} V_k - EV_1|| \to 0$$

in probability.

Next, these results will be used to show that for identically distributed random elements in separable normed linear spaces the weak law of large numbers in the weak linear topology is necessary and sufficient for the weak law of large numbers in the norm topology. Thus, weak laws of large numbers are available for weakly uncorrelated random elements. The following theorem of Taylor (1972) will now be obtained by embedding the separable normed linear space isomorphically in the Banach space C[0,1] and by applying Theorem 4.1.3.

Theorem 4.1.5 Let X be a separable normed linear space and let $\{V_n\}$ be a sequence of identically distributed random elements in X such that $E||V_1|| < \infty$ and EV_1 exists. For each $f \in X^*$, the weak law of large numbers holds for the sequence $\{f(V_n)\}$ if and only if

$$||\frac{1}{n} \sum_{k=1}^{n} V_k - EV_1|| \to 0$$

in probability.

Proof: It is sufficient to prove the "only if" part since convergence in the norm topology implies convergence in the weak linear topology. Let $\hat{X}$ be the completion of X. Since $\hat{X}$ is isometric to a subspace of C[0,1] [Marti (1969), page 67], there exists a one-to one, bicontinuous, linear function h from X into C[0,1].

By Lemma 2.3.1 $\{h(V_n)\}$ is a sequence of identically distributed random elements in C[0,1] and $E||h(V_1)|| \le ||h||E||V_1|| < \infty$. Let $g \in C[0,1]^*$, then

$$\frac{1}{n} \sum_{k=1}^{n} g(h(V_k)) = \frac{1}{n} \sum_{k=1}^{n} (h^*g)(V_k) \to$$

$$E[(h^*g)(V_1)] = E[g(h(V_1)] \qquad (4.1.16)$$

where h* is the adjoint function for h and maps C[0,1]* into X*. Thus, for each $g \in C[0,1]^*$ the weak law of large numbers holds for the sequence $\{g(h(V_n))\}$. The space C[0,1] has a Schauder basis (see Example 1.3.2). Thus, by Theorem 4.1.3

$$||\frac{1}{n} \sum_{k=1}^{n} h(V_k) - Eh(V_1)|| \to 0$$

in probability since (4.1.16) holds for each $g \in C[0,1]^*$ and in particular for each coordinate functional. Next, $Eh(V_1) = h(EV_1)$ by Theorem 2.3.7 (iv). Therefore,

$$||\frac{1}{n} \sum_{k=1}^{n} V_k - EV_1|| \to 0$$

in probability since h is one-to-one, bicontinuous and linear. ///

The weak law of large numbers for weakly uncorrelated random elements easily follows as a corollary to Theorem 4.1.5.

<u>Corollary 4.1.6</u> If X is a separable normed linear space and if $\{V_n\}$ is a sequence of identically distributed, weakly uncorrelated random elements in X such that $E||V_1|| < \infty$ and EV_1 exists, then

$$||\frac{1}{n} \sum_{k=1}^{n} V_k - EV_1|| \to 0$$

in probability.

It is interesting to note that the weak law of large numbers in the weak linear topology is sufficient for the weak law of large numbers in the strong (norm) topology but that this is not true for the strong law of large numbers. Beck and Warren (1974) constructed random elements in the separable Banach space c_0 (rf. (2) Section 1.1) which were identically distributed, weakly orthogonal ($E[f(V_\alpha)f(V_\beta)] = 0$ for all $f \in c_0^* = \ell^1$) and uniformly bounded in norm, but which did not satisfy the strong law of large numbers. Hence, convergence with probability one is not obtainable in either Theorem 4.1.3 or Theorem 4.1.5 even though the strong law of large numbers holds for identically distributed, uncorrelated random variables (see Theorem 3.1.2). Moreover, the same example was produced by Beck and Warren (to appear) in a separable, reflexive, and uniformly convex Banach space. However, a strong law of large numbers for weakly orthogonal random elements is available in the following theorem by Beck and Warren (1974).

<u>Theorem 4.1.7</u> If X is a Banach space which has a separable dual space X* and if $\{V_n\}$ is a sequence of strictly stationary, weakly orthogonal in X with $E||V_1||^2 < \infty$ and $EV_1 = 0$, then

$$||\frac{1}{n} \sum_{k=1}^{n} V_k|| \to 0$$

with probability one.

The assumption of identically distributed random elements $\{V_n\}$ can not be relaxed in these laws of large numbers by simply imposing bounds on the moments of $\{||V_n||\}$. In particular, Theorems 3.1.1, 3.1.2, 3.1.3, and 3.1.4 for random variables do not extend directly to random elements in separable normed linear spaces. Example 4.1.1 [Beck (1963), page 32] will provide a counterexample and will be used repeatedly to illustrate where other plausible extensions of laws of large numbers to separable normed linear spaces do not hold.

<u>Example 4.1.1</u> Let $X = \ell^1 = \{x \in R^\infty: \ ||x|| = \sum_{n=1}^{\infty} |x_n| < \infty\}$ and let δ^n denote the element having 1 for its nth term and 0

elsewhere. Let A_n be an independent sequence of random variables defined by $A_n = \pm 1$ each with probability 1/2, and define $V_n = A_n \delta^n$. Clearly, $\{V_n\}$ is an independent sequence of random elements in ℓ^1 with $||V_n|| \equiv 1$ and $EV_n = 0$ for each n. But,

$$||\frac{1}{n} \sum_{k=1}^{n} V_k|| = ||\frac{1}{n}(A_1, A_2, \ldots, A_n, 0, \ldots)||$$

$$= \frac{1}{n} \sum_{k=1}^{n} |A_k| \equiv 1$$

for each n. Hence,

$$\frac{1}{n} \sum_{k=1}^{n} V_k \not\to 0$$

in any mode including convergence in distribution. ///

Example 4.1.1 clearly indicates a major problem in obtaining stochastic convergence results for infinite dimensional spaces. In the next section distributional conditions on the random elements are used to obtain laws of large numbers. In Section 4.3 convexity (or lack of convexity) is related to laws of large numbers as B(Beck)-convexity is considered. The last section of this chapter examines characterizations of the Banach spaces which are necessary and sufficient for strong laws of large numbers.

4.2 DISTRIBUTIONAL CONDITIONS AND LAWS OF LARGE NUMBERS

In this section distributional conditions are used to obtain laws of large numbers. Since there are no additional restrictions on the separable normed linear spaces, Example 4.1.1 indicates that these distributional conditions will be rather restrictive. However, for particular pplications the results of this section can be very useful and may be easily applied (see Example 4.2.1). The distribution al condition of tightness provides several laws of large numbers, but these will be included in the more general results for weighted sums as corollaries in Chapter V.

Beck and Giesy (1970) studied P-uniform convergence in normed linear spaces and as a consequence obtained strong laws of large numbers for normed linear spaces by suitably restricting the random elements. Since they considered strongly measurable random elements, their results will be stated for separable normed linear spaces where the less restrictive definition of a random element coincides with the definition of a strongly measurable random element (see Property 2.1.4).

Recall that $\beta||V||$ denotes the essential supremum of the random variable $||V||$ when V is a random element. Also, $\sigma^2(V)$ and $\sigma(V)$ denotes respectively the variance and standard deviation of the random element V.

Theorem 4.2.1 [Beck and Giesy (1970)] If X is a separable normed linear and if $\{V_n\}$ is a sequence of independent random elements in X such that $EV_n = 0$ for each n and such that

$$\frac{1}{n} \sum_{k=1}^{n} \beta||V_k|| \to 0 \tag{4.2.1}$$

as $n \to \infty$, then

$$||\frac{1}{n} \sum_{k=1}^{n} V_k|| \to 0$$

with probability one.

Theorem 4.2.2 Beck and Giesy (1970) If X is a separable normed linear space and if $\{V_n\}$ is a sequence of independent random elements in X such that $EV_n = 0$ for each n and such that

$$\sum_{n=1}^{\infty} \frac{\sigma^2(V_n)}{n^2} < \infty \quad \text{and} \quad \lim_{n\to\infty} \frac{1}{n} \sum_{k=1}^{n} \sigma(V_k) = 0, \tag{4.2.2}$$

then

$$||\frac{1}{n}\sum_{k=1}^{n} V_k|| \to 0$$

with probability one.

Beck and Giesy (1970) further concluded that if the restriction on $\{\beta||V_n||\}$ in Condition (4.2.1) is weakened or if the restrictions on $\{\sigma(V_n)\}$ in Condition 4.2.2 are weakened, then the results no longer hold for all separable normed linear spaces. Taylor and Padgett (1974) obtained strong laws of large numbers by using product sequences of random variables and random elements.

Suppose that the random element V_1 in X and random variables $\{A_n\}$ satisfy the following conditions:

$$E||V_1||^{\frac{2r}{r-1}} < \infty \text{ and } \sum_{n=1}^{\infty} (E[|A_n|^{2r}])^{\frac{1}{r}}/n^2 < \infty \tag{4.2.3}$$

for some $r > 1$, and

$$E||V_1||^{\frac{s}{s-1}} < \infty \text{ and } \frac{1}{n}\sum_{k=1}^{n}(E[|A_k|^{s}])^{\frac{1}{s}} \le L \tag{4.2.4}$$

for all n and for some $s > 1$ and $L > 0$.

<u>Theorem 4.2.3 [Taylor and Padgett (1974)]</u> Let X be a separable normed linear space. Let $\{A_n\}$ be a sequence of real-valued random variables and let $\{V_n\}$ be a sequence of identically distributed random elements in X such that Conditions (4.2.3) and (4.2.4) hold. If $\{A_nV_n\}$ is a sequence of independent random elements and if $E(A_nV_n) = E(A_1V_1)$ for each n, then

$$||\frac{1}{n}\sum_{k=1}^{n} A_kV_k - E(A_1V_1)|| \to 0$$

with probability one.

A more useful form of Theorem 4.2.3 for applications is obtained by requiring the following conditions to hold:

$$E||V_1||^2 < \infty \text{ and } \sum_{n=1}^{\infty} (\beta|A_n|)^2/n^2 < \infty \tag{4.2.5}$$

and

$$\frac{1}{n} \sum_{k=1}^{n} \beta|A_k| \le L \tag{4.2.6}$$

for all n where $L > 0$ [Taylor and Padgett (1974)]. The following example gives an application where the Cesáro boundedness of Condition (4.2.6) is satisfied but where the Cesáro convergence to zero of Condition (4.2.2) does not hold. The example will also show that identically distributed random elements are often easily obtained from random elements which are not identically distributed.

Example 4.2.1 Let $\{Z_n\}$ be a sequence of separable Brownian motion processes on $[0,1]$ such that $\{\sigma_n^2 = E[Z_n^2(1)]\}$ satisfies the condition that $\{\frac{1}{n} \sum_{k=1}^{n} \sigma_k\}$ is a bounded sequence and that $\sum_{n=1}^{\infty} \sigma_n^2/n^2 < \infty$. Each Z_n can be regarded as a random element in $C[0,1]$, and Theorem 4.2.3 can be applied by letting $Z_n = \sigma_n V_n$ if the $\{Z_n\}$ are independent. By construction the $\{V_n\}$ are identically distributed since $1 = E[V_n^2(1)]$ (Property 2.3.5). Conditions (4.2.5) and (4.2.6) are satisfied since $A_n = \sigma_n$ and $E||V_1||^4 < \infty$. Hence,

$$\sup_{t\in[0,1]} \left|\frac{1}{n} \sum_{k=1}^{n} Z_k(t)\right| \to 0$$

with probability one since $E(Z_n)$ is the zero function on $[0,1]$ for each n. ///

Theorem 4.2.3 can be viewed as a randomly weighted sum of random elements. Results on randomly weighted sums are considered in Chapter VI. However, it is important to note that independence is required for the product sequence $\{A_n V_n\}$ but not for the identically distributed random elements $\{V_n\}$. For a non-trivial example, let $\Omega = \{a,b,c,d\}$, A = all subsets of Ω, and $P[\{\omega\}] = 1/4$ for each $\omega \in \Omega$. Define

$$V_1(\omega) = I_{\{a,d\}}(\omega) - I_{\{b,c\}}(\omega) = -V_2(\omega)$$

and

$$A_1(\omega) \equiv 1 \quad \text{and } A_2(\omega) = I_{\{a,b\}}(\omega) - I_{\{c,d\}}(\omega).$$

Note that V_1 and V_2 are dependent but that A_1V_1 and A_2V_2 are independent.

The product sequence technique also allows the identical distribution condition of Theorems 4.1.3 and 4.1.5 to be slightly relaxed.

Theorem 4.2.4 [Taylor (1972)] Let X be a separable normed linear space and let $\{V_n\}$ be a sequence of identically distributed random elements in X such that $E||V_1||^{\frac{s}{s-1}} < \infty$ for some $s > 1$. Also let $\{A_n\}$ be a sequence of random variables such that for each n

$$\frac{1}{n} \sum_{k=1}^{n} (E[|A_k|^s])^{\frac{1}{s}} \leq L$$

where L is a positive constant, and let $E(A_nV_n) = E(A_1V_1)$ for each n. For each $f \in X^*$ the weak law of large numbers holds for the sequence $\{f(A_nV_n)\}$ if and only if

$$||\frac{1}{n} \sum_{k=1}^{n} A_kV_k - E(A_1V_1)|| \to 0$$

in probability.

If X has a Schauder basis such that $\{||U_n||\}$ is a bounded sequence for the partial sum operators $\{U_n\}$, then the weak law of large numbers holding in each coordinate is necessary and sufficien for the weak law of large numbers in Theorem 4.2.4.

4.3 BECK'S CONVEXITY AND THE STRONG LAW OF LARGE NUMBERS

In this section the strong law of large numbers for independent random variables with uniformly bounded variances will be extended to separable normed linear spaces which satisfy Beck's convexity condition (see Definition 4.3.1 below). The convexity condition is a necessary and sufficient condition for this extension. However, Example 4.3.1 will show that the strong law of large numbers for random variables which satisfy Kolmogorov's condition on the variances (see Theorem 3.2.4) does not extend to separable normed linear spaces even if the convexity condition is satisfied. Also included in this section will be a discussion of Giesy's (1965) results for normed linear spaces with this convexity condition and other implications.

A normed linear space X is said to be uniformly convex if for every $\varepsilon > 0$ there exists a $\delta > 0$ such that $||x|| \le 1$, $||y|| \le 1$, and $||x + y|| > 2 - 2\delta$ implies that $||x - y|| < \varepsilon$ for all $x, y \in X$. Beck (1958) extended the strong law of large numbers for random variables with uniformly bounded variances to separable, uniformly convex Banach spaces. This result will follow as a corollary to Theorem 4.3.1.

Definition 4.3.1 A normed linear space X is said to be convex of type (B) if there is an integer $t > 0$ and an $\varepsilon > 0$ such that for all $x_1, \ldots, x_t \in X$ with $||x_i|| \le 1$, $i = 1, \ldots, t$, then

$$||\pm x_1 \pm x_2 \pm \ldots \pm x_t|| < t\,(1 - \varepsilon) \qquad (4.3.1)$$

for some choice of + and - signs.

Giesy (1965) extensively studied the convexity property of normed linear spaces which is given in Definition 4.2.1. Finite-dimensional normed linear spaces, uniformly convex normed linear spaces (and hence the L^p-spaces and ℓ^p-spaces, $1 < p < \infty$), and inner product spaces are convex of type (B). Examples of normed linear spaces which are not convex of type (B) include ℓ^1, ℓ^∞, and c_0. Giesy (1965) also characterized type (B) convex spaces by conditions on their first and second conjugate spaces and on factor spaces. One interesting geometric characterization of spaces which are not convex of type (B) is that they must have isomorphic copies of finite dimensional ℓ^1 for arbitrary finite dimension. Recall that ℓ^1 in Example 4.1.1 had this property. More detailed results on spaces which are convex of type (B) may be found in Giesy (1965). The following several pages will be used to reproduce Beck's (1963) strong law of large numbers for random elements in normed linear spaces which are convex of type (B) and whose variances are uniformly bounded.

<u>Theorem 4.3.1</u> If X is a separable normed linear space which is convex of type (B) and if $\{V_n\}$ is a sequence of independent random elements in X such that $EV_n = 0$ and $E||V_n||^2 \leq M$ for all n where M is a positive constant, then

$$||\frac{1}{n} \textstyle\sum_{k=1}^{n} V_k|| \to 0$$

with probability one.

Since every normed linear space is isomorphic to a dense subspace of a Banach space, it suffices to prove Theorem 4.3.1 for separable Banach spaces which are convex of type (B). The proof which will be given is essentially the proof contained in Beck (1963)

<u>Definition 4.3.2</u> A random element V in a normed linear space X is said to be <u>symmetric</u> if there exists a measure-preserving function ϕ of Ω into Ω (that is, $P[\phi^{-1}(B)] = P[B]$ for each $B \in A$) such that $P[V \circ \phi = -V] = 1$.

Recall that $\beta||V||$ denotes the essential supremum of the random variable $||V||$. For a sequence of random elements in X, define

$$c\{V_n\} = \beta(\limsup_n ||\tfrac{1}{n} \textstyle\sum_{k=1}^n V_k||). \qquad (4.3.2)$$

A sequence of random elements is said to be of <u>type (A)</u> if they are bounded in norm by 1, have the zero element as their expected values, and are symmetric and independent. Finally, define

$$C(X) = \sup \{c\{V_n\}: \ \{V_n\} \text{is of type (A) in X}\} \qquad (4.3.3)$$

where the supremum is taken over all sequences of random elements where are of type (A) in X. Note that $0 \le C(X) \le 1$ since for any sequence of random elements $\{V_n\}$ of type (A)

$$\begin{aligned} c\{V_n\} &= \beta(\limsup_n ||\tfrac{1}{n} \textstyle\sum_{k=1}^n V_k||) \\ &\le \beta(\limsup_n \tfrac{1}{n} \textstyle\sum_{k=1}^n ||V_k||) \\ &\le \beta(\limsup_n 1) = 1. \end{aligned}$$

In part (a) of the proof of Theorem 4.3.1, it will be shown that $C(X) = 0$ when X is convex of type (B).

<u>Proof of Theorem 4.2.1 - Part (a)</u>: Let X be a separable Banach space which is convex of type (B) and suppose that $C(X) = C \neq 0$. Hence, for $\eta > 0$ there exists a sequence of random elements $\{W_n\}$ in X which are of type (A) and such that $c\{W_n\} > C - \eta$. Define

$$Z_n = \frac{W_{tn}+W_{tn-1}+\ldots+W_{tn-t+1}}{t} \qquad (4.3.4)$$

where t is the positive integer which is given by the type B convexity of X. Without loss of generality, it can be assumed that the sequence of measure-preserving functions $\{\phi_n\}$ which correspond to the random elements $\{W_n\}$ have the property that

$$P[W_m \circ \phi_n = W_m] = 1 \qquad (4.3.5)$$

for $m \neq n$ and

$$P[W_n \circ \phi_n = -W_n] = 1 \qquad (4.3.6)$$

for each n, otherwise, consider the infinite product of the original probability space (Ω,A,P) and let the random element W_n be identified with the nth coordinate. Note,

$$E(Z_n) = \frac{E(W_{tn})+E(W_{tn-1})+\ldots+E(W_{tn-t+1})}{t} = 0,$$

$$||Z_n|| \leq \frac{||W_{tn}||+||W_{tn-1}||+\ldots+||W_{tn-t+1}||}{t} \leq 1,$$

and the random elements $\{Z_n\}$ are independent. For the symmetry of each Z_n, let $\phi_{Z_n} = \phi_{tn} \circ \phi_{tn-1} \circ \ldots \circ \phi_{tn-t+1}$. Thus, $\{Z_n\}$ is of type (A) and

$$c\{Z_n\} = \beta(\limsup_n ||\tfrac{1}{n} \textstyle\sum_{k=1}^n Z_k||)$$

$$= \beta(\limsup_n ||\tfrac{1}{n} \textstyle\sum_{k=1}^n W_k||) = c\{W_n\} \qquad (4.3.7)$$

since $||W_n|| \leq 1$ for each n. After possibly eliminating a countable union of null sets from Ω and possibly considering the equivalent countable infinite product of the probability space (Ω,A,P) with itself, Conditions (4.3.5) and (4.3.6) hold pointwise (for each $\omega \in \Omega$). For each n there are 2^t possible measure-preserving functions on Ω which can be formed by taking compositions of the elements of the subsets of $\{\phi_{tn}, \phi_{tn-1}, \ldots, \phi_{tn-t+1}\}$. Moreover, for each $\omega \in \Omega$ one of these compositions, Φ_ω, has the property that

$$||\textstyle\sum_{k=tn-t+1}^{nt} W_k(\Phi_\omega(\omega))|| < t(1-\varepsilon) \qquad (4.3.8)$$

since X is convex of type (B). Label these 2^t functions as as $\Phi_1,\ldots,\Phi_{2^t}$. Thus, for each n and each $\omega \in \Omega$

$$\sum_{r=1}^{2^t} ||\sum_{k=tn-t+1}^{nt} W_k(\Phi_r(\omega))||$$
$$< t(2^t - 1) + t(1 - \varepsilon) = t(2^t - \varepsilon). \qquad (4.3.9)$$

But since the functions $\{\Phi_r\}$ are measure-preserving, (4.3.9) implies that

$$2^t E(||\sum_{k=tn-t+1}^{nt} W_k||) = E(\sum_{r=1}^{2^t} ||\sum_{k=tn-t+1}^{tn} W_k \circ \Phi_r||)$$
$$\leq t(2^t - \varepsilon). \qquad (4.3.10)$$

From (4.3.10) it follows that for all n

$$E||Z_n|| = \frac{1}{t} E||\sum_{k=tn-t+1}^{tn} W_k||$$
$$\leq \frac{1}{t} \frac{t(2^t-\varepsilon)}{2^t} = 1 - \frac{\varepsilon}{2^t}. \qquad (4.3.11)$$

Let $s > \frac{1}{\eta^3}$, and for each n define

$$Y_n = \frac{Z_{sn}+Z_{sn-1}+\ldots+Z_{sn-s+1}}{s}. \qquad (4.3.12)$$

Again, it follows that $\{Y_n\}$ is of type (A) and that $c\{Z_n\} = c\{Y_n\}$. Similar to (4.3.11) it can be shown that $E||Z_n||^2 < 1$, and hence $\mathrm{Var}\,(||Z_n||) < 1$. But since $\{||Z_n||\}$ is a sequence of independent random variables,

$$\mathrm{Var}(\sum_{k=sn-s+1}^{sn} ||\frac{Z_k}{s}||) < \frac{1}{s}. \qquad (4.3.13)$$

Thus, it follows from (4.3.11) that for each n

$$P[||Y_n|| > 1 - \frac{\varepsilon}{2^t} + \eta] \leq P[\sum_{k=sn-s+1}^{sn} ||\frac{Z_k}{s}|| > 1 - \frac{\varepsilon}{2^t} + \eta]$$
$$\leq P[\sum_{k=sn-s+1}^{sn}(||\frac{Z_k}{s}|| - E||\frac{Z_k}{s}||) > \eta]$$
$$\leq \frac{\frac{1}{s}}{\eta^2} < \eta. \qquad (4.3.14)$$

For each n define

$$P_n = Y_n I_{[||Y_n|| \le 1 - \frac{\varepsilon}{2^t} + \eta]} \tag{4.3.15}$$

and

$$Q_n = Y_n - P_n. \tag{4.3.16}$$

The random elements $\{P_n\}$ are independent and uniformly bounded in norm by $1 - \frac{\varepsilon}{2^t} + \eta$. The symmetry of the random elements follows from the symmetry of $\{Y_n\}$ and hence $E(P_n) = 0$ for each n. Thus, $\{P_n\}$ is of type (A) and $c\{P_n\} \le C(1 - \frac{\varepsilon}{2^t} + \eta)$. The random variables $\{||Q_n||\}$ are independent, $||Q_n|| \le 1$, and $P[||Q_n|| = 0] > 1 - \eta$ by (4.3.14). Thus, $E||Q_n|| < \eta$ for all n, and by the strong law of large numbers $c\{Q_n\} \le c\{||Q_n||\} \le \eta$. Moreover,

$$c\{Y_n\} \le c\{P_n\} + c\{Q_n\} \le c\{P_n\} + \eta. \tag{4.3.17}$$

Thus, by the initial assumption and by construction,

$$\begin{aligned} C - \eta < c\{W_n\} &= c\{Z_n\} = c\{Y_n\} \\ &\le c\{P_n\} + \eta \\ &\le C(1 - \frac{\varepsilon}{2^t} + \eta) + \eta. \end{aligned} \tag{4.3.18}$$

A contradiction follows since $0 < C \le 1$ implies that

$$\frac{\varepsilon}{2^t} < \frac{3\eta}{C}$$

for all $\eta > 0$. Hence, $C = 0$, and the proof of part (a) is completed. ///

It is easy to see that the proof of part (a) remains valid if the random elements are uniformly bounded in norm by some arbitrary constant. In part (b) of the proof, Theorem 4.3.1 is proved for independent, symmetric random elements $\{V_n\}$ such that $\text{Var}(V_n) \le M$ for all n.

Proof of Theorem 4.3.1 - Part (b): Let X be a separable Banach space which is convex of type (B) and let $\{V_n\}$ be a sequence of independent, symmetric random elements such that Var $(V_n) \le M$ for all n where M is a positive constant. Again, it can be assumed that $M = 1$ without loss of generality. Let m be an arbitrary positive integer and for each n define the random elements Y_n and Z_n by

$$Y_n = V_n I_{[||V_n|| \le m]} \text{ and } Z_n = V_n - Y_n. \tag{4.3.19}$$

Using the fact that $EV_n = 0$,

$$\begin{aligned} E||Z_n|| &= \frac{1}{m} E(m||Z_n||) \le \frac{1}{m} E||Z_n||^2 \\ &\le \frac{1}{m} E||V_n||^2 = \frac{1}{m} \text{Var } (V_n) \le \frac{1}{m} \end{aligned} \tag{4.3.20}$$

for each n. Thus, by the strong law of large numbers for the random variables $\{||Z_n||\}$ and by (4.3.20),

$$c\{Z_n\} \le c\{||Z_n||\} \le \frac{1}{m}.$$

From part (a) of the proof, $c\{Y_n\} = 0$, and hence

$$c\{V_n\} \le c\{Y_n\} + c\{Z_n\} \le \frac{1}{m}.$$

Thus, $c\{V_n\} = 0$. ///

Part (c) of the proof will drop the symmetry condition and hence will complete the proof.

Proof of Theorem 4.3.1 - Part (c): Again, let X be a separable Banach space which is convex of type (B) and let $\{V_n\}$ be a sequence of independent random elements in X such that $E(V_n) = 0$ and

$\mathrm{Var}\,(V_n) \le M$ for all n where M is a positive constant. Consider the probability space $(\Omega,A,P) \times (\Omega,A,P)$. Define a sequence of random elements $\{Y_n\}$ in X by

$$Y_n((\omega_1,\omega_2)) = V_n(\omega_1) - V_n(\omega_2)$$

for each $(\omega_1,\omega_2) \in \Omega \times \Omega$ and for each n. The random elements $\{Y_n\}$ are independent, have expected value 0, and are symmetric (to show symmetry use $\phi(\omega_1,\omega_2) = (\omega_2,\omega_1)$ for each n). In addition,

$$\mathrm{Var}\,(Y_n) = E||Y_n||^2 \le 2E||V_n||^2 + 2E||V_n||^2 \le 4M.$$

Hence, by part (b) of the proof

$$\frac{1}{n}\sum_{k=1}^{n} V_k(\omega_1) - \frac{1}{n}\sum_{k=1}^{n} V_k(\omega_2) = \frac{1}{n}\sum_{k=1}^{n} Y_k(\omega_1,\omega_2) \to 0 \tag{4.3.21}$$

with probability one.

Since (4.3.21) defines an equivalence relation, there exists a set $\Omega_0 \in \Omega$ with $P(\Omega_0) = 1$ such that (4.3.21) holds for each $\omega_1,\omega_2 \in \Omega_0$. The proof will be complete once it is shown that

$$||\frac{1}{n}\sum_{k=1}^{n} V_k(\omega_0)|| \to 0$$

for each $\omega_0 \in \Omega_0$. Let $\varepsilon > 0$ be given, let $\omega_0 \in \Omega$ and define

$$E_n = \{\omega \in \Omega: \ ||\frac{1}{n}\sum_{k=1}^{n} V_k(\omega) - \frac{1}{n}\sum_{k=1}^{n} V_k(\omega_0)|| < \varepsilon\}. \tag{4.3.22}$$

Since $P(E_n) \to 1$ as $n \to \infty$, there exists an N such that $P(E_n) > 1/2$ whenever $n > N$. For each continuous linear functional $f \in X^*$, $\{f(V_n)\}$ is a sequence of independent random variables with expected values equal to zero and $\mathrm{Var}\,(f(V_n)) \le ||f||^2\,\mathrm{Var}\,(V_n) \le ||f||^2 M$. For each $f \in X^*$ such that $||f|| \le 1$ and for $n > \frac{2M}{2}$,

$$P[|\frac{1}{n}\sum_{k=1}^{n} f(V_k)| > \varepsilon] \le \frac{1}{(\varepsilon n)^2}\sum_{k=1}^{n} \text{Var}(f(V_n)) < \frac{1}{2}. \quad (4.3.23)$$

Let $D_{n,f}$ denote the set

$$D_{n,f} = \{\omega: |\frac{1}{n}\sum_{k=1}^{n} f(V_k(\omega))| \le \varepsilon\}.$$

For $n \ge \max\{N, \frac{2M}{\varepsilon^2}\}$

$$P[E_n \cap D_{n,f}] = P(E_n) + P(D_{n,f}) - P[E_n \cup D_{n,f}]$$
$$> \frac{1}{2} + \frac{1}{2} - 1 = 0$$

from (4.3.23) and (4.3.22). Thus, for any $f \in X^*$ such that $||f|| \le 1$ and for any $n > \max\{N, \frac{2M}{\varepsilon^2}\}$, there exists an element $\omega_n \in E_n \cap D_{n,f}$ and hence

$$|\frac{1}{n}\sum_{k=1}^{n} f(V_k(\omega_0))| \le |\frac{1}{n}\sum_{k=1}^{n} f(V_k(\omega_0) - V_k(\omega_n))| + |\frac{1}{n}\sum_{k=1}^{n} f(V_k(\omega_n))|$$
$$\le ||\frac{1}{n}\sum_{k=1}^{n} (V_k(\omega_0) - V_k(\omega_n))|| + \varepsilon$$
$$< 2\varepsilon .$$

Thus, by the Hahn-Banach Theorem (Corollary 1.2.3)

$$||\frac{1}{n}\sum_{k=1}^{n} V_k(\omega_0)|| \le 2\varepsilon$$

whenever $n > \max\{N, \frac{2M}{\varepsilon^2}\}$, and the proof is completed. ///

In addition to Theorem 4.3.1, Beck (1963) showed that convexity of type (B) is necessary to obtain the strong law of large numbers for independent random elements with zero expected values and bounded variances. More precisely, if X is not convex of type (B), then there exists a sequence of independent random elements $\{V_n\}$ with $E(V_n) = 0$ and $\text{Var}(V_n) \le M$ for all n and such that $||\frac{1}{n}\sum_{k=1}^{n} V_k||$ does not converge to zero with probability one. Example 4.3.1 [Beck (1963), Example 15] will show that Kolmogorov's condition on the variances [rf. Theorem 3.1.4] is not sufficient for the strong law of large numbers even in separable normed linear spaces which satisfies Beck's convexity condition.

Example 4.3.1 Let p be a real number such that $1 < p < 2$. Recall that ℓ^p denotes the separable Banach space

$$\ell^p = \{x \in R^\infty: \quad ||x|| = (\sum |x_n|^p)^{\frac{1}{p}} < \infty\}$$

and that δ^n denotes the element having one for its nth coordinate and 0 in the other coordinates. Let q be a real number such that $1 - \frac{1}{p} < q < \frac{1}{2}$. Let $\{A_n\}$ be a sequence of independent random variables defined by $A_n = \pm n^q$ each with probability $\frac{1}{2}$, and define $V_n = A_n \delta^n$ for each n. Thus, $\{V_n\}$ is a sequence of independent random elements in ℓ^p such that $E(V_n) = 0$ for each n and

$$\sum_{n=1}^\infty \frac{Var(V_n)}{n^2} = \sum_{n=1}^\infty \frac{n^{2q}}{n^2} = \sum_{n=1}^\infty \frac{1}{n^{2-2q}} < \infty$$

since $2 - 2q > 1$. But, for n even

$$||\frac{1}{n} \sum_{k=1}^n V_k||^p = \sum_{k=1}^n \frac{k^{pq}}{n^p}$$

$$> \sum_{k=[\frac{n}{2}]}^n \frac{k^{pq}}{n^p} > \sum_{k=[\frac{n}{2}]}^n \frac{(\frac{n}{2})^{pq}}{n^p}$$

$$= \sum_{k=[\frac{n}{2}]}^n \frac{n^{pq-p}}{2^{pq}} > \frac{n^{pq+1-p}}{2^{pq+1}}$$

which goes to ∞ since $1 - \frac{1}{p} < q$ implies that $pq + 1 - p > 0$. ///

Since ℓ^p $(p > 1)$ is also a uniformly convex, reflexive Banach space, Example 4.3.1 also provides a counterexample to the extension of Theorem 3.1.4 to either uniformly convex spaces or reflexive spaces. It was earlier conjectured that all B-convex spaces were reflexive. However, James (1974) presented an example of a non-reflexive B-convex space. The material in this section is only the start of a large area of current research. The next section will develop one aspect of this area, the G_α condition, and discuss the type p spaces, but first the following two theorems of Giesy (to appear) are listed.

Theorem 4.3.2 Let X be a Banach space whose dual space X* is separable. Let $\{V_n\}$ be a sequence of independent random elements with $EV_n = 0$ for all n. If

$$\sum_{n=1}^{\infty} \frac{Var(V_n)}{n^2} < \infty$$

and

$$\{\frac{1}{n} \sum_{k=1}^{n} \sigma(V_k): \quad n \geq 1\}$$

is a bounded sequence, then $\{V_n\}$ satisfies the weak topology strong law of large numbers (WTSLLN), that is, there exists $\Omega_1 \subset \Omega$ such that $P(\Omega_1) = 1$ be such that either

$$f(\frac{1}{n} \sum_{k=1}^{n} V_k(\omega)) \to 0$$

for each $\omega \in \Omega_1$ and $f \in X^*$.

Theorem 4.3.3 Let X be a Banach space which is not B-convex, and let $\{s_k\} \subset R^+$ be such that either

$$\sum_{n=1}^{\infty} \frac{s_n^2}{n^2} = \infty \text{ or } \frac{1}{n} \sum_{k=1}^{n} s_k \to \infty.$$

Then, there exists a sequence of independent random elements $\{V_n\}$ with $EV_n = 0$ and $\sigma(V_n) \leq s_n$ for all n which fail to satisfy the WTSLLN.

4.4 OTHER GEOMETRIC CONDITIONS AND THE STRONG LAW OF LARGE NUMBERS

In this section geometric conditions are imposed on the Banach spaces to obtain strong laws of large numbers. As indicated in the introduction of the chapter, not all results in this rapidly developing area will be presented. The main goal of this section will be the presentation of Woyczynski's (1973) strong laws of large numbers for independent random elements in "G_α" spaces. Also, the strong law of large numbers for independent random elements in "type p" spaces will be listed, and a brief comparison of the different geometric conditions will be given.

The G_α ($\alpha = 1$) was introduced by Fortet and Mourier (1955) and was motivated by the inequality between the moment of a sum of independent random elements and the sum of the moments.

Definition 4.4.1 A Banach space X is said to satisfy the condition G_α for some α, $0 < \alpha \leq 1$, if there exists a mapping $G\colon X \to X^*$ such that

$$(G_\alpha^{I})\colon \quad ||G(x)|| = ||x||^\alpha,$$
$$(G_\alpha^{II})\colon \quad G(x)x = ||x||^{1+\alpha}, \text{ and}$$
$$(G_\alpha^{III})\colon \quad ||G(x) - G(y)|| \leq A||x - y||^\alpha$$

for all $x,y \in X$ and some positive constant A.

Theorem 4.4.1 Let X be a separable Banach space which is G_α for some $0 < \alpha \leq 1$ and let $\{V_1,\ldots,V_n\}$ be independent random elements in X such that $EV_k = 0$ and $E||V_k||^{1+\alpha} < \infty$ for each $k = 1,2,\ldots n$. Then,

$$E(||V_1+\ldots+ V_n||^{1+\alpha}) \leq A \sum_{k=1}^n E||V_k||^{1+\alpha}$$

where A is the positive constant in (G_α^{III}).

Proof: From (G_α^{II})

$$||V_1 + \ldots + V_n||^{1+\alpha} = G(V_1 + \ldots + V_n)(V_1 + \ldots + V_n)$$
$$= \sum_{k=1}^n G(V_1 + \ldots + V_n)V_k \qquad (4.4.1)$$

pointwise (for each $\omega \in \Omega$). Let $T_j = \sum_{j \neq k} V_k$. Then $V_1 + \ldots + V_n = T_j + V_j$ for each j, or

$$(V_1 + \ldots + V_n) - T_j = V_j. \qquad (4.4.2)$$

Define V_j^* pointwise for each j by

$$G(V_1 + \ldots + V_n) - G(T_j) = V_j^*. \qquad (4.4.3)$$

Since X is separable and G is continuous (G_α^{III}), V_j^* is a strongly measurable random element in X*. Moreover, from (4.4.2), (4.4.3), and (G_α^{III})

$$||V_j^*|| \leq A||V_j||^\alpha. \qquad (4.4.4)$$

Hence,

$$\begin{aligned} E||V_1 + \ldots + V_n||^{1+\alpha} &= E[\textstyle\sum_{j=1}^n G(V_1 + \ldots + V_n)V_j] \\ &= E[\textstyle\sum_{j=1}^n (G(T_j)V_j + V_j^* V_j)] \\ &\leq \textstyle\sum_{j=1}^n E[G(T_j)](EV_j) + E[||V_j^*||\ ||V_j||] \end{aligned} \qquad (4.4.5)$$

But, $E[V_j] = 0$ for each $j = 1,\ldots,n$, and

$$||V_j^*|| \leq A||V_j||^\alpha$$

pointwise implies that $E[||V_j^*||\ ||V_j||] \leq E[A||V_j||^{1+\alpha}]$. Thus, (4.4.5) becomes

$$E||V_1 + \ldots + V_n||^{1+\alpha} \leq A \textstyle\sum_{j=1}^n E||V_j||^{1+\alpha}. \qquad ///$$

Woyczynski (1973) provided a Lipschitzian condition for conditions (G_α^I), (G_α^{II}), and (G_α^{III}). Relating this to the (Gateaux derivative) gradient of the norm and the smoothness of the space X, it follows that L^p and ℓ^p $(p \geq 2)$ are G_α for any α, $0 < \alpha \leq 1$. Also,

$\ell^{1+\alpha}$ and $L^{1+\alpha}$ are G_α for $0 < \alpha \le 1$, but $\ell^{1+\alpha}$ and $L^{1+\alpha}$ are not G_β for $\beta > \alpha$. Note that Hilbert spaces are G_1 with constant $A = 1$ and identity mapping G.

The following theorem from Woyczynski (1973) for convergence of series provides the key step in obtaining the strong law of large numbers.

<u>Theorem 4.4.2</u> Let X be a separable Banach space which is G_α for some $0 < \alpha \le 1$. Let $\{V_n\}$ be a sequence of independent random elements in X such that $EV_n = 0$ for each n. If

$$\sum_{j=1}^{\infty} E\phi_0||V_j|| < \infty \tag{4.4.6}$$

where $\phi_0(t) = t^{1+\alpha}$ for $0 \le t \le 1$ and $\phi_0(t) = t$ for $t \ge 1$, then $\sum_{j=1}^{\infty} V_j$ converges with probability one.

<u>Proof</u>: Define

$$U_j = V_j I_{[||V_j|| \le 1]} \quad \text{and} \quad W_j = V_j I_{[||V_j|| > 1]}. \tag{4.4.7}$$

Note that $V_j = U_j + W_j$ for each j and that both $\{U_j\}$ and $\{W_j\}$ are independent sequences of random elements in X. Next, for each m and n

$$E||\sum_{j=n}^{m} W_j|| \le \sum_{j=n}^{m} E||W_j|| \le \sum_{j=n}^{m} E\phi_0||V_j||, \tag{4.4.8}$$

and hence

$$E||\sum_{j=1}^{m} W_j|| \text{ converges}$$

as $m \to \infty$ by a Cauchy argument. Since convergence in the mean implies convergence in probability, Ito and Nisio's (1968) result for independent random elements (cf. Section 4.5) provides that

$$\sum_{j=1}^{\infty} W_j \quad \text{converges}$$

with probability one. From Theorem 4.4.1, (4.4.8), and Jensen's inequality,

$$E||\sum_{j=n}^{m} U_j||^{1+\alpha} \le E(||\sum_{j=n}^{m} (U_j-EU_j)|| + ||\sum_{j=n}^{m} EU_j||)^{1+\alpha}$$

$$\le 2^{1+\alpha}(A\sum_{j=n}^{m} E||U_j-EU_j||^{1+\alpha} + ||\sum_{j=n}^{m} EW_j||^{1+\alpha})$$

$$\le 2^{1+\alpha}[A2^{1+\alpha}(\sum_{j=n}^{m} E||U_j||^{1+\alpha} + \sum_{j=n}^{m}||EU_j||^{1+\alpha})+(E||\sum_{j=n}^{m}W_j||)^{1+\alpha}]$$

$$\le 2^{1+\alpha}[A2^{2+\alpha}\sum_{j=n}^{m} E\phi_0||V_j|| + (\sum_{j=n}^{m} E\phi_0||V_j||)^{1+\alpha}]$$

for each n and m. Thus, $E||\sum_{j=1}^{m} U_j||$ is Cauchy, and hence converges. Hence,

$$\sum_{j=1}^{\infty} U_j \quad \text{converges}$$

with probability one. ///

While several more results on the convergence of series of independent random elements in G_α spaces are in Woyczynski (1973), Theorem 4.4.2 and its Corollary are sufficient for the strong law of large numbers since a version of Kronecker's lemma is available for locally convex linear topological spaces.

Corollary to Theorem 4.4.2 Let X be a separable Banach space which is G_α for some $0 < \alpha \le 1$. Let $\{V_n\}$ be a sequence of independent random elements in X such that $EV_n = 0$ for each n. If ϕ_n: $R^+ \to R^+$, n = 1,2,..., are continuous and such that $\phi_n(t)/t$ and $t^{1+\alpha}/\phi_n(t)$ are non-decreasing, then for each $\{\alpha_n\} \subset R^+$ the convergence of

$$\sum_{n=1}^{\infty} \frac{E\phi_n||V_n||}{\phi_n(\alpha_n)}$$

implies that

$$\sum_{n=1}^{\infty} \frac{V_n}{\alpha_n}$$

converges with probability one.

Lemma 4.4.3 Kronecker's lemma Let $\{x_k\}$ be a sequence in a locally convex linear topological space X and let $\{\alpha_k\}$ be a sequence of positive numbers tending to infinity. Then

$$\sum_{k=1}^{\infty} \frac{1}{\alpha_k} x_k \quad \text{convergent}$$

implies that

$$\frac{1}{\alpha_n} \sum_{k=1}^{n} x_k \to 0.$$

Theorem 4.4.4 Let X be a separable Banach space which is G_α for some $0 < \alpha \leq 1$ and let $\phi\colon\ R^+ \to R^+$ be a continuous function such that $\phi(t)/t$ and $t^{1+\alpha}/\phi(t)$ are non-decreasing. If $\{V_n\}$ are independent random elements in X such that $EV_n = 0$ for each n and

$$\sum_{n=1}^{\infty} \frac{1}{\phi(n)} E\phi||V_n|| < \infty, \tag{4.4.9}$$

then

$$\frac{1}{n} \sum_{k=1}^{n} V_k \to 0$$

with probability one. In particular, an analogue of the Kolmogorov's strong law of large numbers ($\phi(t) = t^2$) holds for random elements in ℓ^p if and only if $p \geq 2$.

The convergence of the series (4.4.9) implies that

$$\sum_{j=1}^{\infty} \frac{V_j}{j} \quad \text{converges}$$

with probability one by the Corollary to Theorem 4.4.2. Thus, Kronecker's lemma yields

$$\frac{1}{n} \sum_{j=1}^{n} V_j \to 0$$

with probability one. Hence, Theorem 4.4.4 follows. The final corollary on G_α results follows from the characterizations in Woyczynski (1973).

Corollary 4.4.5 In ℓ^p, $1 < p < \infty$, the convergence of

$$\sum_{n=1}^{\infty} \frac{E||V_n||^{1+\alpha}}{n^{1+\alpha}} \qquad 0 < \alpha \le 1$$

implies the strong law of large numbers for independent random elements $\{V_n\}$ with $EV_n = 0$ if and only if $\ell^p \in G_\alpha$, that is $p \ge 1 + \alpha$.

Let $\{E_n\}$ be a Bernoulli sequence, that is, $\{E_n\}$ are independent random variables with $P[E_n = +1] = \frac{1}{2} = P[E_n = -1]$ for each n. Let $X^\infty = \prod_{n=1}^{\infty} X$ and define

$$C(X) = \{(x_n) \in X^\infty\colon \sum_{n=1}^{\infty} E_n x_n \text{ converges in probability}\}.$$

Then a separable Banach space X is said to be of type p, $1 \le p \le 2$, if and only if there exists $A \in R^+$ such that

$$E||\sum_{n=1}^{\infty} E_n x_n||^p \le A \sum_{n=1}^{\infty} ||x_n||^p$$

for all $(x_n) \in C(X)$. The following two results and accompanying remarks are from Hoffmann-Jørgensen and Pisier (1976).

Theorem 4.4.6 Let $1 \le p \le 2$, then the following four statements are equivalent:

(i) The separable Banach space X is of type p.

(ii) There exists $A \in R^+$ such that

$$E||\sum_{k=1}^{n} V_k||^p \le A \sum_{k=1}^{n} E||V_k||^p$$

for all independent random elements $V_1, \ldots, V_n$ with mean 0 and finite pth moments.

(iii) The strong law of large numbers,

$$\frac{1}{n} \sum_{k=1}^{n} V_k \to 0 \text{ with probability one,}$$

holds for all sequences of independent random elements $\{V_n\}$ such that $EV_n = 0$ for all n and such that Chung's condition holds,

$$\sum_{n=1}^{\infty} \frac{E||V_n||^p}{n^p} < \infty.$$

(iv) If

$$\sum_{n=1}^{\infty} \frac{||x_n||^p}{n^p} < \infty$$

and $\{E_n\}$ is a Bernoulli sequence, then

$$\frac{1}{n} \sum_{k=1}^{n} E_k x_k \to 0$$

in probability.

Pisier (1974) showed that X is B-convex if and only if X is of type p for some $p > 1$. By Theorem 4.4.6 every G_α space is of type $(1 + \alpha)$. However, the G_α spaces are much smoother than type $(1 + \alpha)$ spaces. In fact, X is a G_α space, $0 < \alpha \le 1$, if and only if X is uniformly $(1 + \alpha)$ smooth, that is, the modulus of smoothness

$$\rho(t) = \sup \{\tfrac{1}{2}(||x+y||+||x-y||-2): \quad ||x|| = 1 \text{ and } ||y|| = t\}$$

satisfies $\rho(t) = O(t^{1+\alpha})$ as $t \to 0$. Thus, G_α spaces are superreflexive, but James (1974) has given an example of a B-convex space (and hence type $(1 + \alpha)$ for some α) which is not reflexive and not G_α.

<u>Theorem 4.4.7</u> Let X be a separable Banach space and let $\{V_n\}$ be independent random elements in X with $EV_n = 0$ for all n and satisfying for each $\varepsilon > 0$ there exists a compact set $K \subset X$ so that

$$E(||V_n|| I_{[V_n \notin K]}) \le \varepsilon \qquad \text{for all } n,$$

then

$$\lim_{n\to\infty} E[\frac{1}{n}||\sum_{k=1}^{n} V_k||] = 0.$$

Since convergence in the mean implies convergence in probability, Theorem 4.4.7 yields a weak law of large numbers. The condition involving the compact set K resembles the condition of tightness which was used by Taylor and Wei (to appear) to obtain strong and weak laws of large numbers. Since every B-convex space is of type p for some p, $1 \le p \le 2$, and Chung's condition appears less restrictive than the condition of uniformly bounded variances, Example 4.3.1 appears to present a contradiction. But note in Example 4.3.1 that $q > 1 - \frac{1}{p}$ and hence

$$\sum_{n=1}^{\infty} \frac{E||V_n||^p}{n^p} = \sum_{n=1}^{\infty} \frac{n^{qp}}{n^p} > \sum_{n=1}^{\infty} \frac{n^{p-1}}{n^p} = \sum_{n=1}^{\infty} \frac{1}{n} = \infty.$$

4.5 OTHER CONVERGENCE RESULTS AND EXTENSIONS TO FRECHET SPACES

Although Chapter IV is concerned with laws of large numbers for separable normed linear spaces, some related results will be listed in this section which are rates of convergence or which are in spaces which are not normed linear spaces. The laws of large numbers which will be discussed in this section are concerned with convergence in the rth mean. A result for sums of independent random elements will also be stated. Finally, the extensions of the laws of large numbers to certain Fréchet spaces will be outlined.

Mourier (1956) proved that if X is a separable Banach space and if $\{V_n\}$ is a sequence of independent, identically distributed random elements in X such that $E||V_1||^r < \infty$, $1 \le r < \infty$, then

$$E(||\tfrac{1}{n} \textstyle\sum_{k=1}^{n} V_k - EV_1||^r) \to 0.$$

In addition, if X* is separable and if $r \ge 2$, then there exists a positive number ρ such that

$$E(||\tfrac{1}{n} \textstyle\sum_{k=1}^{n} V_k||^r) \ge \rho n^{-r/2}$$

for all n. Also, a reverse inequality may be obtained by restricting the Banach space to be a G_1 space [see Mourier (1956)].

For separable Banach spaces which are convex of type (B), Giesy (1965) obtained the following result which gives an upper bound on the rate of convergence for a law of large numbers of the Mourier type stated above: Let $1 \le p < q \le \infty$ with $2 \le q$. If X is a separable Banach space which is convex of type (B) and if $\{V_n\}$ is a sequence of independent random elements in X with $EV_n = 0$ and $(E||V_n||^q)^{1/q} \le M$ for all n $[\beta||V_n|| \le M$ for $q = \infty]$, then there exist real numbers $b_n = b_n(p,q,t,\varepsilon)$ such that $b_n \to 0$ and

$$(E||\frac{1}{n} \sum_{k=1}^{n} V_k||^p)^{1/p} \le Mb_n$$

for all n.

Alf (preprint) extended the results on the convergence rates of Lai (1974) and Heyde and Rohatgi (1967) to independent random elements in separable Banach spaces.

Keulbs (1976 & 1976) has obtained Banach space versions of the law of the iterated logarithm in addition to several other related convergence results.

Ito and Nisio (1968) proved the following result concerning sums of independent random elements: Let $\{V_n\}$ be a sequence of independent random elements in a separable Banach space and let $S_n = V_1 + \ldots + V_n$. Then the following conditions are equivalent:

(i) S_n converges with probability one,

(ii) S_n converges in probability, and

(iii) the distribution of S_n converges in the Prohorov metric [see Billingsley (1968)].

Recall that the Fréchet space F in (10) of Section 1.1 has a metric which is given by the Fréchet combination of a family of seminorms $\{p_k\}$. For each k the space F with seminorm p_k is a seminormed linear space which is denoted by F_k, and the topology of each seminormed linear space F_k is weaker than the metric topology of F. Also, convergence (either in probability or with probability one) in all of the seminormed linear spaces F_k, $k = 1,2,\ldots$, is equivalent to convergence in the metric topology of F. For each $x \in F_k$, let $||\hat{x}||_k = p_k(x)$ where $\hat{x}$ denotes the embedded element in the quotient space F_k/N_k with $N_k = \{y \in F_k\colon\ p_k(y) = 0\}$. Thus, if the corresponding moment conditions are defined in terms of the seminorms $\{p_k\}$, then several of the laws of large numbers of this chapter extend easily to F since the embeddings preserve the probabilistic properties. Additional details on these extensions are available in Chapter VI of Padgett and Taylor (1973).

The strong law of large numbers for independent, identically distributed random elements in F was obtained by Ahmad (1965) using a different method of proof, but the above embedding technique is simpler and allows further results. For example, strong laws of large numbers in F which are not necessarily for identically distributed random elements may be obtained by using the results of Section 4.2.

For the weak law of large numbers to hold for identically distributed random elements in F, it is necessary and sufficient for the weak law of large numbers to hold in the weak linear topology of F. This result follows from Theorem 4.1.5 and the indicated embedding technique since f is a continuous linear functional on F if f is a continuous linear functional on each F_k. Thus, the weak law of large numbers also holds for weakly uncorrelated random elements in F.

4.6 PROBLEMS

4.1 Let $V_1,\ldots,V_t$ be random elements in a separable normed linear space X. Show that for each $\varepsilon > 0$

$$P[\textstyle\sum_{i=1}^{t} ||V_i|| > \varepsilon] \le \sum_{i=1}^{t} P[||V_i|| > \frac{\varepsilon}{t}].$$

4.2 Prove or Disprove Let $\{V_n\}$ be random variables such that $V_n \to 0$ with probability one. Then

$$\frac{1}{n}\sum_{k=1}^{n} V_k \to 0 \text{ with probability one.}$$

4.3 Prove or Disprove Let $\{V_n\}$ be random variables such that $V_n \to 0$ in probability. Then

$$\frac{1}{n}\sum_{k=1}^{n} V_k \to 0 \text{ in probability.}$$

4.4 Show that every uniformly convex space is convex of type (B) by finding the appropriate t and ε.

4.5 Verify that $c\{Z_n\} = c\{W_n\}$ in (4.3.7).

4.6 Show that the inequality in Theorem 4.4.1 does not hold for $\ell^1 = \{x \in R^\infty: \|x\| = \sum|x_n| < \infty\}$ for any $\alpha > 0$ and hence ℓ^1 is not G_α.

4.7 Show that every separable Banach space is of type 1.

4.8 Conjecture: Define independent random elements $\{V_n\}$ in c, the space of convergent sequences, $\|V_n\| \equiv 1$ and $EV_n = 0$ for all n, but where

$$\left\|\frac{1}{n}\sum_{k=1}^{n} V_k\right\| \equiv 1 \text{ for all } n.$$

(Neither the author nor any of his colleagues have resolved the conjecture, and there is a split opinion on its validity.)

CHAPTER V

CONVERGENCE OF WEIGHTED SUMS IN NORMED LINEAR SPACES

5.0 INTRODUCTION

In this chapter the convergence of the weighted sums

$$S_n = \sum_{k=1}^{n} a_{nk}V_k \text{ or } = \sum_{k=1}^{\infty} a_{nk}V_k$$

are obtained under various conditions on the weights $\{a_{nk}\}$, the random elements $\{V_n\}$, and the normed linear spaces. Laws of large numbers will be special cases of convergence results for weighted sums of random elements. In particular, in Section 5.2 very general and powerful convergence results are proved for weighted sums of tight random elements with uniformly bounded rth moments ($r > 1$). These results yields new laws of large numbers without requiring geometric conditions on the spaces and supplements the results of Sections 4.1 and 4.2. Also, in Section 5.2 comparisons of these results with previous work are discussed, and several examples are presented to disprove other plausible conjectures.

In Section 5.1 several results for the convergence of weighted sums are obtained using the identical distributions of the random elements, moment conditions on the random elements, and restrictions on the weights. For identically distributed random elements in a separable normed linear space, convergence in probability in the weak linear topology is necessary and sufficient for convergence in probability for the weighted sum in the norm topology. Next, let $\lim_{n\to\infty} a_{nk} = 0$ for every k, $\max_k |a_{nk}| = \mathcal{O}(n^{-\alpha})$ for some $\alpha > 0$, $\sum_{k=1}^{\infty} |a_{nk}|$ be uniformly bounded for all n, and $\{V_k\}$ be independent, identically distributed random elements in a separable normed linear space X with $EV_1 = 0$. Then if $E||V_1||^{1+1/\alpha} < \infty$, $S_n \to 0$ in X with probability one. Conditions similar to Rohatgi's dominance in probability are introduced in relaxing the identically distributed condition.

In Section 5.3 geometric conditions of the normed linear spaces are used to obtain additional results. For example, it is shown that (under slightly different conditions on $\{a_{nk}\}$ than those stated above) weighted sums of independent, not necessarily identically distributed, random elements $\{V_k\}$ with $EV_k = 0$ converge with probability one to the zero element if the random elements have uniformly bounded rth absolute moments for some $r > 1$ and X is a normed linear space satisfying Beck's convexity condition. Finally, the type p spaces are used in Section 5.3 to obtain convergence in probability for weighted sums of independent random elements.

5.1 DISTRIBUTIONAL CONDITIONS AND CONVERGENCE OF WEIGHTED SUMS

Recall that a Toeplitz sequence is a double array of real numbers $\{a_{nk}\}$ satisfying

$$\lim_{n\to\infty} a_{nk} = 0 \quad \text{for each } k, \tag{5.1.1}$$

and

$$\sum_{k=1}^{\infty} |a_{nk}| \le C \quad \text{for each } n \tag{5.1.2}$$

where C is some positive constant. Theorem 5.1.1 [Taylor and Padgett, (1975)] will show that convergence in probability in each coordinate of a Schauder basis for a Banach space is a necessary and sufficient condition for weighted sums of identically distributed random elements to converge in probability in the norm topology.

Theorem 5.1.1 Let X be a Banach space which has a Schauder basis $\{b_i\}$, let $\{a_{nk}\}$ be a Toeplitz sequence, and let $\{V_n\}$ be a sequence of identically distributed random elements in X such that $E||V_1|| < \infty$. For each coordinate functional f_i

$$\sum_{k=1}^{n} a_{nk} f_i(V_k - EV_1) \to 0$$

in probability if and only if

$$||\sum_{k=1}^{n} a_{nk}(V_k - EV_1)|| \to 0$$

in probability.

The proof of Theorem 5.1.1 follows in the same manner as the proof of Theorem 4.1.3 and hence will not be presented. In Theorem 5.1.2 the dual space will be used in place of the coordinate functionals, and the proof is obtained by embedding the separable normed linear space isomorphically in C[0,1] (which has a Schauder basis) and by applying Theorem 5.1.1.

Theorem 5.1.2 Let X be a separable normed linear space, let $\{a_{nk}\}$ be a Toeplitz sequence, and let $\{V_n\}$ be a sequence of identically distributed random elements in X such that $E||V_1|| < \infty$ and EV_1 exists. For each continuous linear functional f

$$\sum_{k=1}^{n} a_{nk} f(V_k - EV_1) \to 0$$

in probability if and only if

$$||\sum_{k=1}^{n} a_{nk}(V_k - EV_1)|| \to 0$$

in probability.

If the identically distributed random elements $\{V_n\}$ are weakly uncorrelated and if the condition that $\max_{1 \le k \le n} |a_{nk}| \to 0$ as $n \to \infty$ is assumed [as in Pruitt (1966), Rohatgi (1971), and Padgett and Taylor

(1974)], then for $\varepsilon > 0$

$$P[|\sum_{k=1}^{n} a_{nk} f(V_k - EV_1)| > \varepsilon]$$

$$\leq \frac{1}{\varepsilon^2} \sum_{k=1}^{n} a_{nk}^2 E[f(V_k - EV_1)^2]$$

$$\leq \frac{1}{\varepsilon^2} (\max_{1 \leq k \leq n} |a_{nk}|) \sum_{k=1}^{n} |a_{nk}| Var[f(V_k)]$$

$$\leq \frac{\max_{1 \leq k \leq n} |a_{nk}|}{\varepsilon^2} C \, Var[f(V_1)]. \tag{5.1.3}$$

Thus, by (5.1.3) and Theorem 5.1.2 as $n \to \infty$

$$||\sum_{k=1}^{n} a_{nk}(V_k - EV_1)|| \to 0 \tag{5.1.4}$$

in probability. Moreover, if the identically distributed random elements are independent, then $\max_{1 \leq k \leq n} |a_{nk}| \to 0$ as $n \to \infty$ will yield the convergence in (5.1.4) by [Pruitt's (1966)] Theorem 3.4.5 and Theorem 5.1.2 of this section. If the Banach space has a Schauder basis, then the more general (see comparisons in Section 3.3) coordinate uncorrelation will provide the convergence theory. Also, Theorem 5.1.1 in conjunction with [Pruitt's (1966)] Theorem 3.4.6 will yield convergence in probability of the weighted sums when the random elements are independent in each coordinate of some Schauder basis for the space.

Example 4.1.1 again illustrates some of the futility in trying to obtain convergence results for nonidentically distributed random elements without assuming stringent moment conditions. For example, if

$$\sum_{k=1}^{n} a_{nk} E||V_k - EV_k|| \to 0,$$

then

$$||\sum_{k=1}^{n} a_{nk}(V_k - EV_k)|| \to 0$$

in probability. In attempting to obtain somewhat less trivial convergence results than the one previously listed, Example 4.1.1 provides the counterexample for many plausible extensions of random variable results. However, results can be obtained for classes of random elements which need not be identically distributed by using product sequences of random variables and random elements. The product sequences results are in Taylor and Padgett (1976) and will not be reproduced here.

Next, let

$$\max_k |a_{nk}| = \mathcal{O}(n^{-\alpha}) \quad \text{for some } \alpha > 0. \tag{5.1.5}$$

Theorem 5.1.3 Let $\{V_k\}$ be a sequence of independent, identically distributed random elements in a separable normed linear space X with $EV_1 = 0$ and let $\{a_{nk}\}$ be a Toeplitz sequence which satisfies Condition (5.1.5). If $E||V_1||^{1+1/\alpha} < \infty$, then

$$S_n = \sum_{k=1}^n a_{nk}V_k \to 0$$

with probability one.

Proof: Let $\{b_i\}$ be a Schauder basis for C[0,1], and let h denote the one-to-one bicontinuous linear function from X into C[0,1]. Define $h(V_k) = V_{hk}$. For each coordinate functional of the basis, f_i, the sequence of random variables $\{f_i(V_{hk})\}$ are independent and identically distributed by Lemma 2.3.1.

For each fixed m,

$$\begin{aligned} ||U_m(\textstyle\sum_{k=1}^n a_{nk}V_{hk})|| &= ||\textstyle\sum_{i=1}^m f_i(\sum_{k=1}^n a_{nk}V_{hk})b_i|| \\ &\le \textstyle\sum_{i=1}^m \left|\sum_{k=1}^n a_{nk}f_i(V_{hk})\right| ||b_i|| \to 0 \end{aligned} \tag{5.1.6}$$

with probability one as $n \to \infty$, since by Theorem 3.4.6 and the fact that

$$E|f_i(V_{h1})|^{1+1/\alpha} \leq ||f_i \circ h||^{1+1/\alpha} E||V_1||^{1+1/\alpha} < \infty,$$

it follows that

$$|\textstyle\sum_{k=1}^{n} a_{nk} f_i(V_{hk})| \to 0$$

with probability one.

Now, for each m

$$||Q_m(\textstyle\sum_{k=1}^{n} a_{nk}V_{hk})|| \leq \textstyle\sum_{k=1}^{n} |a_{nk}|[||Q_m(V_{hk})||-E||Q_m(V_{h1})||]$$
$$+ CE||Q_m(V_{h1})||$$

by Condition (5.1.2) and the hypothesis that $\{V_k\}$ are identically distributed. But $\{||Q_m(V_{hk})||-E||Q_m(V_{h1})||\}$ is a sequence of independent, identically distributed random variables with zero means for each m, and by hypothesis $E||Q_m(V_{h1})||^{1+1/\alpha} \leq [||h||(M+1)]^{1+1/\alpha} E||V_1||^{1+1/\alpha} < \infty$, where M is the basis constant. Hence, by Theorem 3.4.6

$$\textstyle\sum_{k=1}^{n} |a_{nk}|[||Q_m(V_{hk})||-E||Q_m(V_{h1})||] \to 0 \qquad (5.1.7)$$

with probability one as $n \to \infty$.

Let Ω_0 be the countable union of the null sets for which (5.1.6) and (5.1.7) do not hold for all $m \geq 1$. Since $||Q_m(V_{h1})|| \to 0$ pointwise, by Lebesgue's dominated convergence theorem, for $\varepsilon > 0$ choose m so that $E||Q_m(V_{h1})|| < \frac{\varepsilon}{3C}$. Thus, on the set $\Omega - \Omega_0$ there exists a positive integer $N(\varepsilon,\omega)$ by (5.1.6) and (5.1.7) such that $n \geq N(\varepsilon,\omega)$ implies

$$||h(S_n(\omega))|| \leq ||U_m(\textstyle\sum_{k=1}^{n} a_{nk}V_{hk}(\omega))|| + ||Q_m(\textstyle\sum_{k=1}^{n} a_{nk}V_{hk}(\omega))||$$
$$< \frac{2\varepsilon}{3} + CE||Q_m(V_{h1})||$$
$$< \varepsilon.$$

Now, since h is one-to-one, bicontinuous and linear from X into C[0,1], $h^{-1}[h(S_n)] = S_n$ converges with probability one to the zero element of X as $n \to \infty$ whenever $h(S_n)$ converges with probability one to the zero element of C[0,1] as $n \to \infty$, completing the proof. ///

In Example 4.1.1 $P[||V_n|| \geq a] \leq P[||V_1|| \geq a]$ for all $a > 0$ and every n. Also, $E||V_1||^{\gamma} = 1 < \infty$ for any $\gamma > 0$. Thus, Rohatgi's condition [rf. (3.4.12)] alone will not be sufficient in trying to relax the condition of identical distributions. However, a result similar to Theorem 3.4.10 for weighted sums of independent, not necessarily identically distributed, random elements $\{V_k\}$ in a Banach space with a Schauder basis may be obtained. Assume that there exist random variables $\{A_i\}$ and $\{A_m\}$ such that for each coordinate functional f_i of the basis, the random variables $\{f_i(V_k)\}$ are uniformly dominated in probability in the sense that

(A) $P[|f_i(V_k)| \geq a] \leq P[|A_i| \geq a]$

for all $a > 0$ and each k, and such that

(B) $E|A_i|^{1+1/\alpha} < \infty$ and $E|A_m|^{1+1/\alpha} < \infty$

for some $\alpha > 0$ and each i and m, and for each $m \geq 1$

(B cont.) $P\left[\,\left|\,||Q_m(V_k)|| - E||Q_m(V_k)||\,\right| \geq a\right] \leq P[|A_m| \geq a]$

for all $a > 0$ and every k.

<u>Theorem 5.1.4 [Padgett and Taylor (1974)]</u> Let $\{V_k\}$ be a sequence of independent random elements in a Banach space X which has a Schauder basis and let $EV_k = 0$ for each k. Let $\{a_{nk}\}$ be a Toeplitz sequence such that $\max |a_{nk}| = \mathcal{O}(n^{-\alpha})$. If there exist random variables $\{A_i\}$ and $\{A_m\}$ satisfying (A) and (B) and $\sup_k E||Q_m(V_k)|| \to 0$ as $m \to \infty$, then

$$S_n = \sum_{k=1}^{n} a_{nk}V_k \to 0$$

with probability one.

Proof: As in the proof of Theorem 5.1.3, for each fixed m,

$$||U_m(\sum_{k=1}^n a_{nk}V_k)|| \le \sum_{i=1}^m \left|\sum_{k=1}^n a_{nk}f_i(V_k)\right| ||b_i|| \to 0 \qquad (5.1.8)$$

with probability one as $n \to \infty$, since for each $i = 1,2,\ldots,m$ $\{f_i(V_k)\}$ is a sequence of independent random variables which by Theorem 3.4.10 and Conditions (A) and (B) implies that

$$|\sum_{k=1}^n a_{nk}f_i(V_k)| \to 0$$

with probability one. Also, for each m, we have

$$||Q_m(\sum_{k=1}^n a_{nk}V_k)|| \le \sum_{k=1}^n |a_{nk}|[||Q_m(V_k)||-E||Q_m(V_k)||]$$

$$+ \sup_k CE||Q_m(V_k)||$$

from Conditions (5.1.2) and (B). But $\{||Q_m(V_k)||-E||Q_m(V_k)||\}$ is a sequence of independent random variables with zero means for each $m \ge 1$ and by Condition (B) and Theorem 3.4.10

$$\sum_{k=1}^n |a_{nk}|[||Q_m(V_k)||-E||Q_m(V_k)||] \to 0 \qquad (5.1.9)$$

with probability one.

Let Ω_0 be the countable union of null sets for which (5.1.8) and (5.1.9) do not hold for all $m \ge 1$. Given $\varepsilon > 0$ we can choose m so that $\sup_k E||Q_m(V_k)|| < \frac{\varepsilon}{3C}$. Hence on the set $\Omega - \Omega_0$ there exist a positive integer $N(\varepsilon,\omega)$ such that whenever $n \ge N(\varepsilon,\omega)$

$$||S_n|| \le ||U_t(\sum_{k=1}^n a_{nk}V_k)|| + ||Q_t(\sum_{k=1}^n a_{nk}V_k)||$$

$$< \frac{2\varepsilon}{3} + \sup_k E||Q_m(V_k)||$$

$$< \varepsilon,$$

and the proof is complete. ///

For the case X = R, Conditions (A) and (B) reduce to Condition (3.4.12) of Rohatgi (1971) and the condition that $E|V|^{1+1/\alpha} < \infty$

where V is the dominating random variable. Condition (B) is needed to insure that the remainder term [see (5.1.9)] approaches 0 with probability one. The original result in Padgett and Taylor (1974) was stated for dominance by a random element V in the Banach space X. However, in light of Lemma 5.2.1 (rf. Section 5.2), the random variables $\{A_i\}$ and $\{A_m\}$ can more easily be constructed under uniformly bounded moment conditions while tightness (or the particular normed linear space) will yield $\sup_k E||Q_m(V_k)|| \to 0$ as $m \to \infty$. Obviously, these results contain the convergence theory for weighted sums of random vectors when X is taken to be R^n. For R^n only Condition A is required to hold for some Schauder basis.

In Theorem 5.1.3 it was required that $E||V_1||^{1+1/\alpha} < \infty$ for some $\alpha > 0$ and $\max_k |a_{nk}| = \mathcal{O}(n^{-\alpha})$. This moment condition can be relaxed by requiring different conditions on the weights $\{a_{nk}\}$, as is shown in the following theorem of Padgett and Taylor (1976).

<u>Theorem 5.1.5</u> Let X be a separable normed linear space and let $\{V_k\}$ be independent, identically distributed random elements in X with $E||V_1|| < \infty$ and $EV_1 = 0$. Let $\{d_{nk}\}$ be an array of real numbers satisfying $\limsup_{n\to\infty} \sum_{k=1}^n d_{nk}^2 < \infty$ and define

$$a_{nk} = \begin{cases} \frac{d_{nk}}{n}, & k = 1,2,\ldots,n \\ 0, & k > n \end{cases}$$

for each $n = 1,2,\ldots$. Then

$$S_n = \sum_{k=1}^n a_{nk} V_k \to 0$$

with probability one.

<u>Proof</u>: Again, it suffices to prove the result for a Banach space which has a Schauder basis. Let $m > 0$ denote the basis constant. For each fixed positive integer m, consider

$||U_m(\sum_{k=1}^n a_{nk}V_k)|| \le \sum_{i=1}^m \left|\sum_{k=1}^n a_{nk}f_i(V_k)\right| \cdot ||b_i||$. For each $i = 1,\ldots,m$, $\{f_i(V_k)\}$ is a sequence of independent, identically distributed random variables with $Ef_i(V_1) = 0$. Also

$$E|f_i(V_1)| \le ||f_i||E||V_1|| < \infty.$$

Hence, by Theorem 3.4.13

$$\sum_{k=1}^n a_{nk}f_i(V_k) \to 0$$

with probability one for each $i = 1,2,\ldots,m$. Hence,

$$||U_m(\sum_{k=1}^n a_{nk}V_k)|| \to 0 \tag{5.1.10}$$

with probability one for each m.

Now, for each fixed m, $\{||Q_m(V_k)|| - E||Q_m(V_k)||\}$ is a sequence of independent, identically distributed random variables with zero means. Also,

$$E\left|\,||Q_m(V_k)|| - E||Q_m(V_k)||\,\right| \le 2(M + 1)E||V_1|| < \infty$$

by hypothesis. Hence, again by Theorem 3.4.13 it follows that for each m

$$\sum_{k=1}^n |a_{nk}|[||Q_m(V_k)|| - E||Q_m(V_k)||] \to 0 \tag{5.1.11}$$

with probability one.

Since $||Q_m(V_1)|| \to 0$ pointwise, $E||Q_m(V_1)|| \to 0$ as $m \to \infty$. Also, by hypothesis, $\limsup_{n\to\infty} \sum_{k=1}^n d_{nk}^2 < \infty$ implies that $\sup_n \sum_{k=1}^n d_{nk}^2 < \infty$. Hence, for $\varepsilon > 0$, m can be chosen so large that

$$E||Q_m(V_1)|| < \frac{\varepsilon}{3(1+\sup_n \sum_{k=1}^n d_{nk}^2)}. \tag{5.1.12}$$

In addition, for all $n \ge 1$

$$\sum_{k=1}^{n} |a_{nk}| = \sum_{k=1}^{n} \frac{|d_{nk}|}{n} \le \sum_{k=1}^{n} \frac{(1+d_{nk}^2)}{n}$$
$$= 1 + \frac{1}{n} \sum_{k=1}^{n} d_{nk}^2$$
$$\le 1 + \sum_{k=1}^{n} d_{nk}^2$$
$$\le 1 + \sup_n \sum_{k=1}^{n} d_{nk}^2 < \infty.$$

Now, let Ω_0 be the countable union (over m) of null sets for which (5.1.10) and (5.1.11) do not hold. Given $\varepsilon > 0$ and $\omega \notin \Omega_0$, let m be sufficiently large so that (5.1.12) holds. Then for this m (5.1.10) and (5.1.11) insures that there exists $N(\varepsilon,\omega)$ such that for all $n \ge N(\varepsilon,\omega)$

$$||S_n(\omega)|| \le ||U_m(\sum_{k=1}^{n} a_{nk} V_k(\omega))|| + ||Q_m(\sum_{k=1}^{n} a_{nk} V_k(\omega))||$$
$$< \frac{2\varepsilon}{3} + \sum_{k=1}^{n} |a_{nk}| E||Q_m(V_1)||$$
$$< \frac{2\varepsilon}{3} + E||Q_m(V_1)||(1 + \sup_n \sum_{k=1}^{n} d_{nk}^2)$$
$$< \varepsilon. \qquad ///$$

The moment condition on $||V_k||$ can be further relaxed by using a result of Stout (1968, page 1557). The following theorem, which requires the finiteness of $E||V_k||^{1/\alpha}$ for some $1 < \alpha \le 2$, is proved in a manner similar to Theorem 5.1.5 and is thus only stated.

Theorem 5.1.6 Let X be a separable normed linear space and let $\{V_k\}$ be independent, identically distributed random elements in X with EV_1 existing and $EV_1 = 0$. Let $\{a_{nk}\}$ be a triangular array of real numbers ($a_{nk} = 0$ if $k > n$). If $E||V_1||^{1/\alpha} < \infty$ for some $1 < \alpha \le 2$,

$$\sum_{k=1}^{n} a_{nk}^2 \le Kn^{-\alpha}, \quad \sum_{k=1}^{n} |a_{nk}|^{2/\alpha} \le Kn^{-\gamma}$$

for some $\gamma > 0$, and $\sum_{k=1}^{n} |a_{nk}| \le K$ for all n, then

$$S_n = \sum_{k=1}^{n} a_{nk}V_k \to 0$$

with probability one.

Note that the sequence

$$a_{nk} = \begin{cases} \frac{1}{n} & k = 1,\ldots,n \\ 0 & k \ge n \end{cases}$$

does not satisfy the conditions of Theorems 5.1.5 and 5.1.6. Hence, strong laws of large numbers do not follow from those two results. This sequence of a_{nk}'s is ruled out in Theorem 5.1.5 in order to obtain the boundedness of $\sum_{k=1}^{n} |a_{nk}|$ so that the remainder term $\sum_{k=1}^{n} |a_{nk}|E||Q_m(V_k)||$ can be made uniformly small. Hence, an interesting question is whether or not strong laws of large numbers can be obtained from weighted sum results which only require finiteness of $E||V_k||^\gamma$ for $\gamma \le 1$. [In general, the answer is no by Example 5.2.3.]

In summary, convergence of weighted sums of independent, but not necessarily identically distributed, random elements in a separable Banach space may be obtained without requiring the conditions (A) and (B) which are troublesome to verify. However, the second moments of the norms of the random elements must be uniformly bounded and the conditions on the weights $\{a_{nk}\}$ are more stringent. For example, the desirable weights $a_{nk} = \frac{1}{n}$ if $k = 1,\ldots,n$ and $a_{nk} = 0$ if $k > n$ did not satisfy the conditions, and hence, a strong law of large numbers could not be obtained. Of course, this is not surprising since the lack of Kolmogorov's strong law of large numbers in Banach spaces was discussed by Beck (1963).

It should be noted that these results may be extended to separable Fréchet spaces whose metric is the Fréchet combination of an increasing total sequence of seminorms using techniques similar to those in Padgett and Taylor (1973, Chapter VI). This extension allows the application of the results to sequences of independent Wiener processes on $[0,\infty)$ and to stochastic processes with countable parameter spaces.

5.2 CONVERGENCE FOR WEIGHTED SUMS OF TIGHT RANDOM ELEMENTS

The classical second moment conditions alone do not yield strong laws for sequences of independent random elements (random variables in topological vector spaces). However, an rth moment condition (where $r > 1$) and the assumption of tightness do lead to strong laws of large numbers as a consequence of results for weighted sums by Taylor and Wei (to appear), and Wei and Taylor (1978b). With the assumption of tight random elements in separable normed linear spaces, convergence of weighted sums results are proved in the section. Tightness of a sequence of random elements has been defined by LeCam (1957), and tightness characterizations of random elements in the function spaces $C[0,1]$ and $D[0,1]$ have been treated by Billingsley (1968). In this section some basic properties of tightness are obtained, and counterexamples are developed for some plausible conjectures. An important property shows that the centered sequence of a sequence of tight random elements with an rth moment condition remains tight. Thus, it can be assumed without loss of generality that the random elements have zero means.

Definition 5.2.1 A random element V in a topological space X is said to tight if for each $\varepsilon > 0$, there exists a compact subset K_ε of X such that $P[V \in K_\varepsilon] > 1 - \varepsilon$. A sequence $\{V_n\}$ of random elements is (uniformly) tight if for each $\varepsilon > 0$, there exists a compact subset K_ε of X such that $P[V_n \in K_\varepsilon] > 1 - \varepsilon$ for all n.

If X is a complete, separable metric space, then each random element in X is tight [Billingsley (1968), page 10]. Hence, any sequence $\{V_n\}$ of identically distributed random elements is tight in a separable, complete metric space since the compact subset chosen for V_1 will serve for all the V_n's. The converse of this assertion is not true by an obvious choice of random variables $\{V_n\}$ where $V_n = 1/n$ with probability 1, $n = 1,2,\ldots$. The sequence $\{V_n\}$ is not identically distributed but is tight by choosing the compact set $[0,1]$.

The following lemma shows that under appropriate conditions, a sequence of tight random elements may be assumed to have zero means without loss of generality.

<u>Lemma 5.2.1</u> Let $\{V_n\}$ be tight random elements in a separable Banach space X. If $\sup_n E||V_n||^r < \infty$ for some $r > 1$, then $\{V_n - EV_n\}$ is tight.

<u>Proof</u>: Let $\Gamma^r = \sup_n E||V_n||^r$. For each positive integer m, let K_m be a compact set in X such that

$$P[V_n \in K_m] > 1 - (1/m\Gamma)^{r/(r-1)} \qquad (5.2.1)$$

for all n. Without loss of generality it may be assumed that K_m is convex and contains zero [Kelley and Namioka (1963), page 113]. Then, for each n, $EV_n I_{[V_n \in K_m]} \in K_m$, and

$$\begin{aligned} ||EV_n - EV_n I_{[V_n \in K_m]}|| &\le E||V_n I_{[V_n \notin K_m]}|| \\ &\le (E||V_n||^r)^{\frac{1}{r}} (P[V_n \notin K_m])^{(r-1)/r} \\ &< 1/m. \end{aligned} \qquad (5.2.2)$$

For each m, K_m is totally bounded and there exist i_m (1/m) -

spheres $\{N(x_i, 1/m)\}$ covering K_m. By Inequality (5.2.2),

$$\{EV_n\} \subset \bigcap_{m=1}^{\infty} \bigcup_{i=1}^{i_m} N(x_i, 2/m). \tag{5.2.3}$$

Since M is complete, the totally bounded set on the right hand side of (5.2.3) has a compact closure, say K. Let $\varepsilon > 0$ be given, and let K_ε be compact such that $P[V_n \in K_\varepsilon] > 1 - \varepsilon$ for all n. Then, by the choice of K and (5.2.3),

$$P[V_n - EV_n \in K_\varepsilon - K] \geq P[V_n \in K_\varepsilon] \geq 1 - \varepsilon \tag{5.2.4}$$

for all n. This completes the proof since $K_\varepsilon - K$ is compact. ///

Lemma 5.2.1 may not be true without the rth moment condition even in a separable Hilbert space.

Example 5.2.1 Let $X = \ell^2$, the space of all square summable sequences. Let $\{V_n\}$ be defined by

$$V_n = \begin{cases} n\delta^n & \text{with probability } 1/n, \\ 0 & \text{with probability } 1 - 1/n, \end{cases}$$

where $\{\delta^n\}$ is the standard basis as in Example 4.1.1. Then $E||V_n|| = 1$ for all n, and $\{V_n\}$ is tight by choosing $\{0\} \cup$ {finite number of $n\delta^n$'s} as the compact set for each $\varepsilon > 0$. Now

$$V_n - EV_n = \begin{cases} (n-1)\delta^n & \text{with probability } 1/n, \\ -\delta^n & \text{with probability } 1 - 1/n. \end{cases}$$

Given $1 > \varepsilon > 0$, any compact set K_ε satisfying

$$P[V_n - EV_n \in K_\varepsilon] > 1 - \varepsilon$$

must contain infinitely many $-\delta^n$'s, which is impossible since then

K_ε would contain a subsequence which is not convergent. Thus, $\{V_n - EV_n\}$ is not tight. ///

Three function spaces in which many stochastic processes take their sample paths are C[0,1], C[0,∞) and D[0,1], the metric space of functions on [0,1] which are right continuous with left limits and endowed with the Skorohod topology. In these spaces, tightness can be characterized by uniform boundedness conditions and uniform equicontinuity conditions (Billingsley 1968 p. 55, p. 125 and Whitt 1970 p. 941) by using the Arzela-Ascoli characterization of compactness. Other characterizations of compactness may be obtained, for example, Mangano (1976) characterized the sequential compactness of certain sequences of Gaussian random elements in C[0,1].

For a sequence of random variables $\{X_n\}$ with uniformly bounded rth moments, $r > 1$, the following lemma concerning their uniform boundedness by a random variable is obtained.

<u>Lemma 5.2.2</u> Let $\{X_n\}$ be a sequence of random variables such that $E|X_n|^r \le \Gamma$ for all n with $\Gamma > 0$ and $r > 1$. Then there exists a random variable X such that

(i) $P[|X_n| \ge a] \le P[|X| \ge a]$ for all n and for all $a \ge 0$,

and

(ii) $E(|X|^{1+1/s}) < \infty$ for $0 < 1/s < r - 1$.

<u>Proof</u>: By the Markov inequality, for $a \ge \Gamma^{\frac{1}{r}}$ and any n

$$P[|X_n| \ge a] \le (E|X_n|^r)/a^r \le \Gamma/a^r = P[|X| \ge a] \quad (5.2.5)$$

where X is the random variable with probability density function

$$f_X(t) = \begin{cases} \dfrac{r\Gamma}{t^{r+1}} & \text{for } t > \Gamma^{\frac{1}{r}} \\ 0 & \text{for } t \le \Gamma^{\frac{1}{r}} . \end{cases}$$

Also, for $a < \Gamma^{\frac{1}{r}}$ and any n,

$$P[|X_n| \geq a] \leq 1 = P[|X| \geq a]. \qquad (5.2.6)$$

Thus, part (i) is proved. For part (ii), observe that

$$E(|X|^{1+1/s}) = \int_{\Gamma^{\frac{1}{r}}}^{\infty} t^{1+1/s}(r\Gamma^{\frac{1}{r}}/t^{r+1})dt$$

$$= \Gamma^{\frac{1}{r}+1} r/(r-1-1/s)\Gamma^{\frac{1}{rs}} < \infty.$$

for $0 < 1/s < r - 1$. ///

Thus, the dominance in probability of Condition (3.4.12) can be accomplished by moment conditions. Next, let X be a separable normed linear space and $\{V_n\}$ be a sequence of random elements in X with zero means and uniformly bounded rth moments ($r > 1$). Let $\{a_{nk}\}$ be a Toeplitz sequence of real numbers. Then by Lemma 5.2.2, $\{||V_n||\}$ is uniformly dominated by a random variable A with $E|A| < \infty$. Since for each n,

$$E(\textstyle\sum_{k=1}^{\infty} ||a_{nk}V_k||) \leq CE|A| < \infty, \qquad (5.2.7)$$

it follows that the series $S_n = \sum_{k=1}^{\infty} a_{nk}V_k$ converges absolutely with probability one. Hence, if the space X is complete, the series S_n converges with probability one.

<u>Theorem 5.2.3</u> Let X be a separable Banach space and let $\{V_n\}$ be tight random elements in X with zero means and where $E||V_n||^r \leq \Gamma$ for all n, for some $r > 1$ and $\Gamma > 0$. Let $\{a_{nk}\}$ be an array of real numbers satisfying $\sum_{k=1}^{\infty} |a_{nk}| \leq C$. Then

$$|\textstyle\sum_{k=1}^{\infty} a_{nk} f(V_k)| \to 0 \text{ in probability}$$

for each $f \in X^*$ if and only if

$$||\textstyle\sum_{k=1}^{\infty} a_{nk}V_k|| \to 0 \text{ in probability.}$$

<u>Proof</u>: Since X can be isometrically embedded in a Banach space with a Schauder basis, it may be assumed without loss of generality that X has a Schauder basis $\{b_t\}$. Assume further that $\Gamma = 1$. The necessity is immediate since convergence in the norm topology implies convergence in the weak linear topology of X.

To prove the sufficiency, let m be the basis constant such that $||U_t|| \leq m$ and $||Q_t|| \leq m + 1$ for each t. For each n and t,

$$\sum_{k=1}^{\infty} a_{nk} V_k = \sum_{k=1}^{\infty} a_{nk} U_t(V_k) + \sum_{k=1}^{\infty} a_{nk} Q_t(V_k).$$

Let $\varepsilon > 0$ be given. For each fixed t,

$$\begin{aligned} P[||U_t(\textstyle\sum_{k=1}^{\infty} a_{nk} V_k)|| > \varepsilon/2] \\ &= P[||\textstyle\sum_{i=1}^{t} f_i(\sum_{k=1}^{\infty} a_{nk} V_k) b_i|| > \varepsilon/2] \\ &\leq P[\textstyle\sum_{i=1}^{t} |f_i(\sum_{k=1}^{\infty} a_{nk} V_k)| \cdot ||b_i|| > \varepsilon/2] \\ &\leq \textstyle\sum_{i=1}^{t} P[|\sum_{k=1}^{\infty} a_{nk} f_i(v_k)| > \varepsilon/2t||b_i||] \\ &\to 0 \end{aligned} \tag{5.2.8}$$

since by hypothesis $|\sum_{k=1}^{\infty} a_{nk} f_i(V_k)| \to 0$ in probability for each i. Let K be a compact set in X such that

$$P[V_n \in K] \geq 1 - (\varepsilon^2/8C(m + 1))^{r/(r-1)} \tag{5.2.9}$$

for all n. Then by Lemma 1.3.3, t can be chosen so that

$$||Q_t(x)|| = ||x - U_t(x)|| < \varepsilon^2/8C \tag{5.2.10}$$

for all $x \in K$. Thus, by Hölder's inequality,

$$\begin{aligned} E||Q_t(V_k)|| &\leq E(||Q_t(V_k)|| I_{[||Q_t(V_k)|| \leq \varepsilon^2/8C]}) \\ &\quad + E(||Q_t(V_k)|| I_{[||Q_t(V_k)|| > \varepsilon^2/8C]}) \end{aligned}$$

$$\le \varepsilon^2/8C + (E||Q_t(V_k)||^r)^{1/r}$$

$$\times (E||I_{[||Q_t(V_k)|| > \varepsilon^2/8C]}||^{r/(r-1)})^{(r-1)/r}$$

$$\le \varepsilon^2/8C + (m + 1)(E||V_k||^r)^{1/r}$$

$$\times (P[||Q_t(V_k)|| > \varepsilon^2/8C])^{(r-1)/r}$$

$$< \varepsilon^2/8C + (m + 1)\varepsilon^2/8C(m + 1)$$

$$= \varepsilon^2/4C \qquad (5.2.11)$$

uniformly in k. Hence, applying Inequality (5.2.11),

$$P[||\textstyle\sum_{k=1}^{\infty} a_{nk}Q_t(V_k)|| > \varepsilon/2]$$

$$\le P[\textstyle\sum_{k=1}^{\infty} |a_{nk}|\ ||Q_t(V_k)|| > \varepsilon/2]$$

$$\le (2/\varepsilon)\textstyle\sum_{k=1}^{\infty} |a_{nk}| E||Q_t(V_k)||$$

$$< (2/\varepsilon)(\varepsilon^2/4C)C$$

$$= \varepsilon/2 \qquad (5.2.12)$$

for each n. For this t, choose N such that

$$P[||U_t(\textstyle\sum_{k=1}^{\infty} a_{nk}V_k)|| > \varepsilon/2] < \varepsilon/2 \qquad (5.2.13)$$

for all $n \ge N$. Then for $n \ge N$, Inequalities (5.2.12) and (5.2.13) imply that

$$P[||\textstyle\sum_{k=1}^{\infty} a_{nk}V_k|| > \varepsilon] \le P[||\textstyle\sum_{k=1}^{\infty} a_{nk}U_t(V_k)|| > \varepsilon/2]$$

$$+ P[||\textstyle\sum_{k=1}^{\infty} a_{nk}Q_t(V_k)|| > \varepsilon/2]$$

$$< \varepsilon.$$

Hence, $\sum_{k=1}^{\infty} a_{nk}V_k \to 0$ in probability. ///

The following example shows that the rth moment condition $(r > 1)$ cannot be reduced to a first moment condition.

<u>Example 5.2.2</u> Let $X = \ell^1$ and $\{\delta^n\}$ be the standard basis. Define a sequence $\{V_n\}$ of independent random elements by

$$V_n = \begin{cases} \pm\sqrt{n}\delta^n & \text{with probability } 1/2\sqrt{n}, \\ 0 & \text{with probability } 1 - 1/\sqrt{n}. \end{cases}$$

Then, $\{V_n\}$ is tight, with zero means and $E||V_n|| = 1$ for all n. Since $(\ell^1)^* = \ell^\infty$, the space of all bounded sequences, for each $f \in (\ell_1)^*$ there exists $\{t_n\} \in \ell^\infty$ and $\Gamma > 0$ with $|t_n| \le \Gamma$ for all n such that

$$f(V_n) = \begin{cases} \pm\sqrt{n}t_n & \text{with probability } 1/2\sqrt{n}, \\ 0 & \text{with probability } 1 - 1/\sqrt{n}. \end{cases}$$

Hence,

$$\begin{aligned} (1/n^2)\textstyle\sum_{k=1}^n \mathrm{Var}(f(V_k)) &= (1/n^2)\textstyle\sum_{k=1}^n \sqrt{k}t_k^2 \\ &\le (\Gamma^2/n^2)n\sqrt{n} \\ &\to 0 \quad \text{as } n \to \infty. \end{aligned} \tag{5.2.14}$$

Thus, $\{f(V_n)\}$ satisfies the weak law of large numbers so that the choice of weights

$$a_{nk} = \begin{cases} 1/n, & k = 1,2,\ldots,n \\ 0, & k = n+1,n+2,\ldots \end{cases}$$

leads to

$$|\textstyle\sum_{k=1}^n a_{nk} f(V_k)| \to 0 \text{ in probability.}$$

But, in this case,

$$||\sum_{k=1}^{n} a_{nk} V_k|| = \sum_{k=1}^{n} a_{nk} ||V_k||$$

$$\to 1 \neq 0 \text{ as } n \to \infty$$

by the weak law of large numbers since

$$(1/n^2)\sum_{k=1}^{n} \mathrm{Var}||V_k|| = (1/n^2)\sum_{k=1}^{n}(\sqrt{k} - 1)$$

$$\leq (1/n^2)(n\sqrt{n} - n)$$

$$\to 0 \text{ as } n \to \infty. \qquad ///$$

In the proof of Theorem 5.2.3, Inequality (5.2.11) is an essential step in obtaining a bound on the infinite dimensional part. A sufficient condition for Inequality (5.2.11) to hold is

$$\lim_{t\to\infty} \sup_{k} E||Q_t(V_k)|| = 0. \qquad (5.2.15)$$

This fact is explicitly stated in the Corollary 5.2.4 below. Thus, Theorem 5.2.3 shows that tightness and the rth moment conditions ($r \geq 1$) lead to the Condition (5.2.15).

Corollary 5.2.4 Let X be a separable Banach space and let $\{V_n\}$ be random elements in X with zero means such that $\lim_{t\to\infty} \sup_{k} E||Q_t(V_k)|| = 0$. Let $\{a_{nk}\}$ be an array such that $\sum |a_{nk}| \leq C$ for each n. For each $f \in X^*$,

$$|\sum_{k=1}^{\infty} a_{nk} f(V_k)| \to 0 \text{ in probability}$$

if and only if

$$||\sum_{k=1}^{\infty} a_{nk} V_k|| \to 0 \text{ in probability.}$$

For a Banach space which has a Schauder basis, to obtain the desired convergence of weighted sums, it is sufficient (and also necessary) to obtain the corresponding convergence property for all coordinate functionals instead of all continuous linear functionals.

Corollary 5.2.5 Let X be a Banach space with a Schauder basis and let $\{V_n\}$ be random elements in X with zero means such that $\lim_{t\to\infty} \sup_k E||Q_t(V_k)|| = 0$. Let $\{a_{nk}\}$ be a Toeplitz sequence. For each coordinate functional f_i,

$$|\sum_{k=1}^{\infty} a_{nk} f_i(V_k)| \to 0 \text{ in probability}$$

if and only if

$$||\sum_{k=1}^{\infty} a_{nk} V_k|| \to 0 \text{ in probability.}$$

Since the double sequence $\{a_{nk}\}$ defined by $a_{nk} = 1/n$ for $k = 1,2,\ldots,n$, and $a_{nk} = 0$ for $k = n+1,\ldots$, is a Toeplitz sequence, Theorem 5.2.3 does provide a necessary and sufficient condition for weak law of large numbers for a sequence of tight zero mean random elements with uniformly bounded rth ($r > 1$) moments in a separable normed linear space. It is interesting to compare this weak law of large numbers with Theorems 4.1.3 and Theorem 4.1.5.

Theorem 5.2.6 Let $\{V_n\}$ be tight random elements in a separable normed linear space such that $E||V_n||^r \le \Gamma$ for each n and for some $r > 1$ and such that EV_n exists for each n. For each $f \in X^*$, the weak law of large numbers holds for $\{f(V_n)\}$ if and only if

$$||\frac{1}{n} \sum_{k=1}^{n} (V_k - EV_k)|| \to 0$$

in probability.

The condition of identical distributions is eliminated by assuming tightness, while an rth ($r > 1$) moment condition is needed instead of the first moment condition. This fact was also reflected in Example 5.2.2. Next, a result is proved on almost sure convergence of weighted sums of independent, tight random elements with uniformly bounded rth moments ($r > 1$) in a separable Banach space. First, the result by Wei and Taylor (1977) will be obtained for random elements which takes their values in a compact subset. Next, the desired result will then be obtained by truncating the random elements to a compact subset and by applying the previous result.

Theorem 5.2.7 Let K be a compact subset of a separable Banach space X. Let $\{V_n\}$ be independent, random elements in X which take their values in K and let $EV_n = 0$ for each n. Let $\{a_{nk}\}$ be a Toeplitz sequence such that $\max_k |a_{nk}| = \mathcal{O}(n^{-\alpha})$ for some $\alpha > 0$. Then

$$S_n = \sum_{k=1}^{\infty} a_{nk} V_k \to 0$$

with probability one.

Proof: It can be assumed that K is convex and symmetric and $0 \in K$ [Rudin (1973), page 72] and that $\sum_{k=1}^{\infty} |a_{nk}| \leq 1$ for each n. In the dual space X* there is a countable set S which separates points of K. Let τ_S be the weakest topology on K making the elements of S continuous. Then for $\{x_n\} \subset K$ $x_n \to 0$ in τ_S if and only if $||x_n|| \to 0$. For each $f \in S$

$$\sum_{k=1}^{\infty} a_{nk} f(V_k) \to 0 \tag{5.2.16}$$

with probability one by Theorem 3.4.6 since $\{f(V_k)\}$ is a sequence of independent, uniformly bounded random variables with zero means. Since K is convex and symmetric

$$\sum_{k=1}^{\infty} a_{nk} V_k(\omega) \in K$$

for each $\omega \in \Omega$ and each n. Since S is countable, (5.2.16) implies that

$$||\sum_{k=1}^{\infty} a_{nk} V_k|| \to 0$$

with probability one. ///

<u>Theorem 5.2.8</u> Let X be a separable Banach space and let $\{V_n\}$ be independent, tight random elements in X with $E||V_n||^r \leq \Gamma$ for all n, where $r > 1$ and $\Gamma > 0$. Let $\{a_{nk}\}$ be a Toeplitz sequence such that $\max_k |a_{nk}| = \mathcal{O}(n^{-s})$ for some $0 < 1/s < r - 1$. Then

$$S_n = \sum_{k=1}^{\infty} a_{nk}(V_k - EV_k) \to 0$$

with probability one.

<u>Proof</u>: By Lemma 5.2.1 it can be assumed that $EV_n = 0$ for all n. Also assume that $\Gamma = 1$ and $\sum_{k=1}^{\infty} |a_{nk}| \leq 1$. Given $\varepsilon > 0$, let

$$\delta = (\varepsilon/4)^{\frac{r}{1-r}}. \tag{5.2.17}$$

Let K be compact (also assume that it is again convex and symmetric and $0 \in K$) such that

$$P[V_n \in K] > 1 - \delta. \tag{5.2.18}$$

Define

$$Y_n = V_n I_{[V_n \in K]} \text{ and } Z_n = V_n - Y_n.$$

Then $\{Y_n - EY_n\}$ takes their values in 2K, and hence

$$||\sum_{k=1}^{\infty} a_{nk}(Y_k - EY_k)|| \to 0 \tag{5.2.19}$$

with probability one by Theorem 5.2.7. By Holder's inequality for each n

$$\begin{aligned} E||Z_n|| &= E(||V_n|| I_{[V_n \notin K]}) \\ &\leq (E||V_n||^r)^{\frac{1}{r}} (P[V_n \notin K])^{\frac{r-1}{r}} \end{aligned}$$

$$\leq 1\ (\delta^{\frac{r-1}{r}}) = \varepsilon/4. \tag{5.2.20}$$

Also, for each n

$$E|\ ||Z_n|| - E||Z_n||\ |^r \leq 2^r. \tag{5.2.21}$$

By Theorem 3.4.10 and Lemma 5.2.2,

$$\sum_{k=1}^{\infty} |a_{nk}|(||Z_k|| - E||Z_k||) \to 0 \tag{5.2.22}$$

with probability one. Since a sequence of $\varepsilon_n \to 0$, $\delta_n \to 0$, and corresponding compact sets K_n could be chosen, a countable number of null sets can be excluded in (5.2.19) and (5.2.22). Thus, for almost all $\omega \in \Omega$, there is an $N(\omega,\varepsilon)$ such that for $n \geq N(\omega,\varepsilon)$,

$$||\sum_{k=1}^{\infty} a_{nk}(Y_k(\omega) - EY_k)|| < \varepsilon/4 \tag{5.2.22}$$

and

$$|\sum_{k=1}^{\infty} a_{nk}(||Z_k(\omega)|| - E||Z_k||)| < \varepsilon/4. \tag{5.2.23}$$

From (5.2.20) and (5.2.23), it follows that

$$\begin{aligned} ||\sum_{k=1}^{\infty} a_{nk} Z_k(\omega)|| &\leq \sum_{k=1}^{\infty} |a_{nk}|(||Z_k(\omega)|| - E||Z_k||) \\ &\quad + \sum_{k=1}^{\infty} |a_{nk}| E||Z_k|| \\ &\leq \varepsilon/4 + \varepsilon/4 = \varepsilon/2 \end{aligned} \tag{5.2.24}$$

for $n \geq N(\varepsilon,\omega)$. Similarly, $EY_k = -EZ_k$, (5.2.20), and (5.2.22) yields

$$||\sum_{k=1}^{\infty} a_{nk} Y_k(\omega)|| < \varepsilon/2 \tag{5.2.25}$$

for all $n \geq N(\varepsilon,\omega)$. Thus, from (5.2.24) and (5.2.25)

$$||\sum_{k=1}^{\infty} a_{nk} V_k(\omega)|| < \varepsilon$$

for all $n \geq N(\varepsilon,\omega)$. ///

Note that Theorem 5.2.8 will provide a strong law of large numbers since the double sequence $\{a_{nk}\}$ defined by $a_{nk} = 1/n$ for $k = 1,2,\ldots,n$, and zero elsewhere, does satisfy the hypothesis of Theorem 5.2.8 when $\max_k |a_{nk}| = \mathcal{O}(n^{-1})$ and $r > 2$. This higher order moment condition is the case as the proof requires the uniform boundedness of $\{||Z_n|| - E||Z_n||\}$ by a random variable in order to apply Theorem 3.4.10. However, the following result of Taylor and Wei (to appear) may be obtained as a corollary which does provide a desired strong law of large numbers for moment conditions with $r > 1$.

Corollary 5.2.9 (SLLN) Let X be a separable Banach space and $\{V_n\}$ be independent tight random elements in X with zero means and $E||V_n||^r \le \Gamma$ for all n and for some $r > 1$ and $\Gamma > 0$. Then

$$||\frac{1}{n} \sum_{k=1}^{n} V_k|| \to 0$$

with probability one.

Proof: In step (5.2.22) of the proof of Theorem 5.2.8, use Chung's (1947) strong law of large numbers to obtain the convergence of

$$\frac{1}{n} \sum_{k=1}^{n} (||Z_k|| - E||Z_k||) \to 0$$

with probability one. The remainder of the proof will be similar to the proof of Theorem 5.2.8. ///

Hoffmann-Jørgenson and Pisier (1976, Theorem 2.4, rf. Theorem 4.4.7 of Section 4.4) stated that if $\{V_n\}$ are independent zero mean random elements in M and satisfy:

For each $\varepsilon > 0$, there exists a compact $K \subset M$ such that

$$E||V_n|| I_{[V_n \notin K]} \le \varepsilon \text{ for all } n, \tag{5.2.26}$$

then

$$\lim_{n\to\infty}\frac{1}{n}E\|\sum_{k=1}^{n}V_k\| = 0. \tag{5.2.27}$$

This result can now be proved by using Lemma 1.3.3 and the basis techniques. In fact, let $\varepsilon > 0$ be given and let $K \subset M$ satisfy (5.2.26). Also, choose t such that $\|x - U_t(x)\| < \varepsilon$ for all $x \in K$. Then

$$\begin{aligned}\frac{1}{n} E\|\sum_{k=1}^{n}V_k\| &\le E\|\frac{1}{n}\sum_{k=1}^{n}V_k I_{[V_k \in K]}\| \\ &\quad + E\|\frac{1}{n}\sum_{k=1}^{n}V_k I_{[V_k \notin K]}\| \\ &\le E\|\frac{1}{n}\sum_{k=1}^{n}U_t(V_k) I_{[V_k \in K]}\| \\ &\quad + E\|\frac{1}{n}\sum_{k=1}^{n}(V_k - U_t(V_k)) I_{[V_k \in K]}\| \\ &\quad + \frac{1}{n}\sum_{k=1}^{n} E\|V_k\| I_{[V_k \notin K]}. \end{aligned} \tag{5.2.28}$$

But

$$E\|\frac{1}{n}\sum_{k=1}^{n}U_t(V_k) I_{[V_k \in K]}\| \to 0 \quad \text{as } n \to \infty \tag{5.2.29}$$

since $\{U_t(V_k) I_{[V_k \in K]}, k = 1,2,\ldots\}$ is a sequence of independent, uniformly bounded, zero mean random elements. The two remaining summands of (5.2.28) are each bounded by ε.

The rth moment condition (some $r > 1$) is essential to the results in this chapter. The following example will show that uniformly bounded first moments are not sufficient for the strong law of large numbers to hold for independent tight random elements. Note that in this example the sequence of random variables is also uniformly integrable.

Example 5.2.3 Let $\{V_n\}$ be independent random variables such that

$$V_n = \begin{cases} n & \text{with probability } 1/(2n\log(n+2)) \\ -n & \text{with probability } 1/(2n\log(n+2)) \\ 0 & \text{otherwise.} \end{cases}$$

Then $EV_n = 0$ and $E|V_n| = 1/\log(n+2) < 1$ for all n. Also, $P[|V_n| \geq n] = 1/n\log(n+2)$. Thus, the random variables $\{V_n\}$ are tight as can be seen by choosing large enough n. Now $\sum_{n=1}^{\infty} P[|V_n| \geq n] = \sum_{n=1}^{\infty} 1/n\ln(n+2) = \infty$ implies by the Borel Lemmas that $|V_n| \geq n$ infinitely often with probability one. That is, there exists $\Omega_0 \subset \Omega$ such that $P(\Omega_0) = 1$ and $|V_n(\omega)| \geq n$ i.o. for all $\omega \in \Omega_0$. Suppose (for a contradiction) that there exists $\Omega_1 \subset \Omega$ such that $P(\Omega_1) = 1$ and $|\frac{1}{n}\sum_{k=1}^{n} V_k(\omega)| \to 0$ for all $\omega \in \Omega_1$. Then, there must exist $\omega \in \Omega_0 \cap \Omega_1$ and $n > 0$ large enough so that $|\frac{1}{n}\sum_{k=1}^{n} V_k(\omega)| < 1/4$, $|\frac{1}{n}\sum_{k=1}^{n-1} V_k(\omega)| < 1/4$, and $\frac{1}{n}|V_n(\omega)| \geq 1$. But then,

$$\begin{aligned} 3/4 &\leq \frac{1}{n}|V_n(\omega)| - |\frac{1}{n}\sum_{k=1}^{n-1} V_k(\omega)| \\ &\leq |\frac{1}{n}\sum_{k=1}^{n} V_k(\omega)| \\ &< 1/4, \end{aligned}$$

a contradiction. ///

The results in this chapter may be extended to a Fréchet space which is a locally convex space with a countable family of seminorms $\{\rho_k\}$ defined on it such that the metric d is defined by

$$d(x,y) = \sum_{k=1}^{\infty} 2^{-k}\rho_k(x - y)/[1 + \rho_k(x - y)].$$

The following theorem and proof can serve as a guide in the

extensions to these Fréchet spaces which facilitate applications in $C[0,\infty)$ and s.

Theorem 5.2.10 Let F be a separable Frechet space and $\{V_n\}$ be independent tight zero mean random elements in F such that $E(\rho_k^r(V_n)) \le \Gamma$ for all n and k, and some $r > 1$. Let $\{a_{ni}\}$ be a Toeplitz sequence of constants. If $\max_i |a_{ni}| = \mathcal{O}(n^{-s})$ for some $0 < 1/s < r - 1$, then

$$\sum_{i=1}^{\infty} a_{ni} V_i \to 0 \quad \text{as } n \to \infty$$

almost surely in the metric topology of F.

Proof: Let F_k be the separable seminormed space with seminorm ρ_k. Since convergence in the metric topology of F is equivalent to convergence in every seminormed space F_k, it suffices to show that for each k, $P[\lim_{n\to\infty} \rho_k(\sum_{i=1}^{\infty} a_{ni} V_i) = 0] = 1$.

For each k there exists a continuous linear functional γ_k from F_k onto the quotient space F_k/N_k, where

$$N_k = \{y \in F_k: \ \rho_k(y) = 0\}.$$

For $x \in F_k$, denote $\gamma_k(x) = \hat{x}$. The space F_k/N_k is a separable normed linear space with norm defined by $||\hat{x}||_k = \rho_k(x)$ and is isomorphic to a dense subset of a separable Banach space Y_k. Hence, there is no ambiguity to identify $\hat{x} \in F_k/N_k$ as an element in Y_k.

Let k be fixed. Since γ_k is continuous, $\{\hat{V}_n\}$ are independent, tight, zero mean random elements in Y_k with $E||\hat{V}_n||_k^r = E(\rho_k^r(V_n)) \le \Gamma$ for all n. Thus, by Theorem 5.2.8,

$$||\sum_{i=1}^{\infty} a_{ni} \hat{V}_i||_k \to 0$$

almost surely. That is,

$$P[\lim_{n\to\infty} \rho_k(\sum_{i=1}^{\infty} a_{ni} V_i) = 0] = 1. \qquad ///$$

5.3 GEOMETRIC CONDITIONS AND CONVERGENCE OF WEIGHTED SUMS

As was illustrated in Chapter IV probabilistic conditions on the sequence of random elements in infinite dimensional spaces provide only a few convergence results unless they are paired with geometric conditions on the space. In this section B-convexity is used to obtain convergence with probability one for weighted sums of independent, not necessarily identically distributed nor tight, random elements whose rth absolute moments are uniformly bounded. Also, convergence in probability is obtained for weighted sums of independent random elements in a type p space using a weaker moment condition. In addition, the sharpness of this result is illustrated by example.

For the following theorem of Taylor and Padgett (1975), let the Toeplitz sequence be nonnegative constants satisfying

$$\sum_{k=1}^{\infty} a_{nk} \leq 1 \tag{5.3.1}$$

and

$$\lim_{n\to\infty} \left[\sum_{k=1}^{n} a_{nk} - n \min_{1\leq k\leq n} \{a_{nk}\}\right] = 0. \tag{5.3.2}$$

Theorem 5.3.1 Let X be a separable normed linear space which is B-convex and let the Toeplitz sequence satisfy (5.3.1) and (5.3.2). Let $\{V_n\}$ be a sequence of independent random elements in X such that $EV_n = 0$ and $E||V_n||^r \leq \Gamma$ for all n and some $\Gamma > 0$ and $r > 1$. If $\max_{1\leq k\leq n} \{a_{nk}\} = \mathcal{O}(n^{-\alpha})$ where $0 < 1/\alpha < r - 1$, then

$$S_n = \sum_{k=1}^{n} a_{nk} V_k \to 0$$

with probability one.

Convergence in probability for Theorem 5.3.1 is easily obtained from Theorem 5.3.2 (see Problem 5.6). As in Beck (1963), the proof of this theorem will be given in three parts. In part (a) the result is obtained by assuming that the random elements are symmetric and uniformly bounded. Part (b) of the proof will consist of replacing the boundedness condition by the condition of uniformly bounded rth moments (some $r > 1$). Finally, the symmetry condition is eliminated in part (c). The essential supremum of the random variable $||V||$ will be denoted by $\beta(||V||)$ throughout the proof.

<u>Proof - Part (a)</u>: Let $\{V_n\}$ be a sequence of independent, symmetric random elements such that $||V_n|| \le \Gamma$ for each n, and let $d_n = \min_{1\le k\le n} \{a_{nk}\}$. It may be assumed without loss of generality that $\Gamma = 1$. Then

$$\begin{aligned} 0 &\le \beta(\limsup_n ||\textstyle\sum_{k=1}^n a_{nk}V_k||) \\ &\le \beta[\limsup_n (||\textstyle\sum_{k=1}^n (a_{nk} - d_n)V_k|| + ||\textstyle\sum_{k=1}^n d_nV_k||)] \\ &\le \beta[\limsup_n \textstyle\sum_{k=1}^n (a_{nk} - d_n)||V_k||] \\ &\quad + \beta[\limsup_n ||d_n \textstyle\sum_{k=1}^n V_k||] \\ &\le \beta[\limsup_n \textstyle\sum_{k=1}^n (a_{nk} - d_n)||V_k||] \\ &\quad + \beta[\limsup_n ||\tfrac{1}{n} \textstyle\sum_{k=1}^n V_k||] \end{aligned} \tag{5.3.3}$$

since $nd_n \le \sum_{k=1}^n a_{nk} \le 1$ by Assumption (5.3.1). But, by Theorem 4.3.1

$$\beta(\limsup_n ||\tfrac{1}{n} \textstyle\sum_{k=1}^n V_k||) = 0. \tag{5.3.4}$$

Also, from the condition that $||V_k|| \le 1$ for all k and assumption (5.2.3),

$$\beta[\limsup_n \textstyle\sum_{k=1}^n (a_{nk} - d_n)||V_k||]$$

$$\leq \limsup_n \sum_{k=1}^n (a_{nk} - d_n) = 0. \qquad (5.3.5)$$

Therefore, from (5.3.3), (5.3.4), and (5.3.5)

$$\beta(\limsup_n ||\sum_{k=1}^n a_{nk}V_k||) = 0,$$

completing part (a) of the proof. ///

Proof - Part (b): Now, let $\{V_n\}$ be a sequence of independent, symmetric random elements in X such that $E||V_n||^r \leq \Gamma$ for all n and some $\Gamma > 0$ and $r > 1$. Without loss of generality, it may be assumed that $\Gamma = 1$.

Let $q > 1$ be a positive integer and define

$$Y_n = V_n \text{ and } Z_n = 0 \text{ if } ||V_n|| \leq q$$

and

$$Y_n = 0 \text{ and } Z_n = V_n \text{ if } ||V_n|| > q.$$

Thus, $\{Z_n\}$ and $\{Y_n\}$ are sequences of independent, symmetric random elements with $||Y_n|| \leq q$ and $EY_n = 0 = EZ_n$ for all n. Hence, by Part (a) of the proof,

$$||\sum_{k=1}^n a_{nk}Y_k|| \to 0 \qquad (5.3.6)$$

with probability one. Also, for each n, by definition of Z_n,

$$\begin{aligned} E||Z_n|| &= (\tfrac{1}{q})^{r-1}E(q^{r-1}||Z_n||) \\ &= (\tfrac{1}{q})^{r-1}(\int_{[||Z_n|| = 0]}q^{r-1}||Z_n||\,dP \\ &\quad + \int_{[||Z_n||>q]}q^{r-1}||Z_n||\,dP) \\ &\leq (\tfrac{1}{q})^{r-1}\int_{[||Z_n||>q]}||Z_n||^r\,dP \\ &\leq (\tfrac{1}{q})^{r-1}E||Z_n||^r \end{aligned}$$

$$\leq (\tfrac{1}{q})^{r-1} E||V_n||^r$$

$$\leq (\tfrac{1}{q})^{r-1} \tag{5.3.7}$$

since $E||V_n||^r \leq 1$ for all n.

Now, the random variables $\{||Z_n|| - E||Z_n||\}$ have mean 0 and since $E||Z_n|| \leq (\frac{1}{q})^{r-1} < 1$, $E(||Z_n||) - ||Z_n|| < 1$ pointwise. Hence, the event $\{\omega \in \Omega: E(||Z_n||) - ||Z_n|| \geq a\} = \phi$ if $a \geq 1$ for all n. Thus, for $a \geq 1$

$$P[|||Z_n|| - E||Z_n||] \geq a] \leq P[||Z_n|| \geq a]$$

$$\leq \frac{E||Z_n||^r}{a^r} \leq \frac{E||V_n||^r}{a^r} .$$

Therefore, by Lemma 5.2.2 and Theorem 3.4.10, as $n \to \infty$

$$\textstyle\sum_{k=1}^{n} a_{nk}(||Z_k|| - E||Z_k||) \to 0$$

with probability one. Hence, there exists an event Ω_1 with $P(\Omega_1) = 1$ such that for every $\omega \in \Omega_1$ and $\delta > 0$ there exists an integer N for which $n \geq N$ implies that

$$\textstyle\sum_{k=1}^{n} a_{nk}E||Z_k|| - \delta < \sum_{k=1}^{n} a_{nk}||Z_k(\omega)||$$

$$< \textstyle\sum_{k=1}^{n} a_{nk}E||Z_k|| + \delta$$

$$< (\tfrac{1}{q})^{r-1} + \delta$$

by (5.3.7) and the assumptions on $\{a_{nk}\}$. Thus,

$$0 \leq \beta(\limsup_n ||\textstyle\sum_{k=1}^{n} a_{nk}Z_k||) \leq (\frac{1}{q})^{r-1}.$$

Combining this result with (5.3.6)

$$0 \leq \beta(\limsup_n ||\textstyle\sum_{k=1}^{n} a_{nk}V_k||)$$

$$\leq \beta(\limsup_n ||\textstyle\sum_{k=1}^{n} a_{nk}Y_k||) + \beta(\limsup_n ||\sum_{k=1}^{n} a_{nk}Z_k||)$$

$$\le (\tfrac{1}{q})^{r-1}.$$

Since $q > 1$ was an arbitrary positive integer, this implies that

$$\beta(\limsup_n ||\textstyle\sum_{k=1}^n a_{nk}V_k||) = 0,$$

completing part (b) of the proof. ///

<u>Proof - Part (c)</u>: This part of the proof proceeds following the steps of the proof of Theorem 4.3.1. However, to obtain the corresponding step that for $\varepsilon > 0$ and sufficiently large n $P[|\sum_{k=1}^n a_{nk}f(V_k)| \ge \varepsilon] < \frac{1}{2}$ for all continuous linear functionals f on X such that $||f|| \le 1$, simultaneously [see (4.3.23) in Section 4.3], it is necessary to observe that the Conditions (5.2.1) and (5.2.2) on the weights $\{a_{nk}\}$ imply that $\max_k |a_{nk}| \to 0$ as $n \to \infty$. Consider two cases: $1 < r < 2$ and $r \ge 2$. If $1 < r < 2$, then $\alpha > 1$ and

$$\begin{aligned} P[|\textstyle\sum_{k=1}^n a_{nk}f(V_k)| \ge \varepsilon] &\le \varepsilon^{-1}\textstyle\sum_{k=1}^n E|a_{nk}f(V_k)| \\ &\le \varepsilon^{-1} \max_{1\le k\le n} |a_{nk}|\textstyle\sum_{k=1}^n ||f||E||V_k|| \\ &\le \varepsilon^{-1} \max_{1\le k\le n} |a_{nk}|\textstyle\sum_{k=1}^n (1 + E||V_k||^r) \\ &\le \varepsilon^{-1} \max_{1\le k\le n} |a_{nk}| n(1 + \Gamma) \end{aligned}$$

since $||f|| \le 1$ and $E||V_k||^r \le \Gamma$ for all k. Thus, from the assumption that $\max_k |a_{nk}| = \mathcal{O}(n^{-\alpha})$ and since $\alpha > 1$, we have for all f such that $||f|| \le 1$

$$P[|\textstyle\sum_{k=1}^n a_{nk}f(V_k)| \ge \varepsilon] \le \varepsilon^{-1}(n^\alpha \max_{1\le k\le n} |a_{nk}|)(1 + \Gamma)n^{1-\alpha}$$

which is less that $\frac{1}{2}$ for n sufficiently large. Now, if $r \ge 2$, using Chebyshev's inequality

$$P[|\sum_{k=1}^{n} a_{nk} f(V_k)| \geq \varepsilon] \leq \varepsilon^{-2} \sum_{k=1}^{n} a_{nk}^2 \, Var[f(V_k)]$$

$$\leq \varepsilon^{-2} \max_{1 \leq k \leq n} |a_{nk}| \sum_{k=1}^{n} a_{nk} ||f||^2 (1 + \Gamma)$$

$$\leq \varepsilon^{-2} (1 + \Gamma) \max_{1 \leq k \leq n} |a_{nk}|. \quad (5.3.8)$$

Hence, for sufficiently large n the right-hand side of (5.3.8) can be made less that $\frac{1}{2}$ simultaneously for all f such that $||f|| \leq 1$. ///

It is important to note that the double sequence $\{a_{nk}\}$ defined by

$$a_{nk} = \begin{cases} \frac{1}{n} & k = 1,2,\ldots,n \\ 0 & k = n+1,\ldots \end{cases}$$

satisfies the Conditions (5.3.1) and (5.3.2). However, this sequence does not satisfy the hypothesis of Theorem 5.3.1 that $\max_{1 \leq k \leq n} \{a_{nk}\} = \mathcal{O}(n^{-\alpha})$ unless $r > 2$, requiring the uniform boundedness of slightly higher order moments than Beck's strong law of large numbers. This is the case since Part (b) of the above proof requires the uniform boundedness of $\{||Z_n|| - E||Z_n||\}$ by the random variable Y in order to use Rohatgi's result, whereas Beck used the strong law of large numbers for random variables to obtain the convergence of $\frac{1}{n} \sum_{k=1}^{n} ||Z_k||$. A number of other possible choices for $\{a_{nk}\}$ are listed in Taylor and Padgett (1975).

Theorem 5.3.2 Let $\{V_n\}$ be independent random elements in a Banach space X of type p, $1 < p \leq 2$. Let V be a random variable such that

$$P[||V_n|| \geq a] \leq P[|V| \geq a] \quad (5.3.9)$$

for all $a > 0$ and that $E|V|^r < \infty$ for some $0 < r < p$. If $r \geq 1$ assume further that $EV_n = 0$ for all n. Let $\{a_{nk}\}$ be a Toeplitz sequence such that $\max_k |a_{nk}| \to 0$ as $n \to \infty$ and $\sum_{k=1}^{\infty} |a_{nk}|^r \le \Gamma$ for all n. Then $S_n = \sum_{k=1}^{n} a_{nk}V_k \to 0$ in probability.

Proof: When $0 < r < 1$, then

$$||\sum_{k=1}^{n} a_{nk}V_k|| \le \sum_{k=1}^{n}|a_{nk}|\ ||V_k|| \to 0$$

in probability by Theorem 3.4.9. The case $r \ge 1$ is more interesting Since $E|V|^r < \infty$,

$$nP[|V| > n^{1/r}] \to 0 \tag{5.3.10}$$

as $n \to \infty$. Let $V_{nk} = a_{nk}V_k I_{[||a_{nk}V_k|| < 1]}$ and $S_{nn} = \sum_{k=1}^{n} V_{nk}$. Then by (5.3.9) and (5.3.10),

$$P[S_{nn} \ne S_n] \le \sum_{k=1}^{n} P[||V_k|| \ge |a_{nk}|^{-1}] \le \sum_{k=1}^{n} P[|V| \ge |a_{nk}|^{-1}]$$

$$= \sum_{k=1}^{n} |a_{nk}|^r(|a_{nk}|^{-r}P[|V| \ge |a_{nk}|^{-1}]) \to 0 \tag{5.3.11}$$

as $n \to \infty$, since $\max_k |a_{nk}| \to 0$ as $n \to \infty$. Thus, it suffices to show that $S_{nn} \to 0$ in probability.

Integration by parts yields that

$$\int_0^T x^p dP[||V_k|| < x]dx = T^p P[||V_k|| < T] - p\int_0^T x^{p-1}P[||V_k|| < x]dx$$

$$= T^p - T^p P[||V_k|| \ge T] - T^p + p\int_0^T x^{p-1}P[||V_k|| \ge x]dx$$

$$\le p\int_0^T x^{p-1}P[|V| \ge x]dx. \tag{5.3.12}$$

Given $\delta > 0$, choose $B > 0$ such that $x \ge B$ implies

$$P[|V| \ge x] \le \delta x^{-r}. \tag{5.3.13}$$

Then, for $T > M(\delta)$

$$T^{r-p}\int_0^T x^{p-1}P[|V| \ge x]dx \le T^{r-p}(\int_0^B x^{p-1}P[|V| \ge x]dx + \delta\int_B^T x^{p-1-r}dx)$$

$$\le T^{r-p}(B^p + \delta T^{p-r}) < 2\delta. \tag{5.3.14}$$

By (5.3.12), (5.3.14), and the definition of type p space,

$$P[||S_{nn}|| \ge \varepsilon] \le \varepsilon^{-p}E||S_{nn}||^p \le \varepsilon^{-p}E||\sum_{k=1}^{n} V_{nk}||^p$$

$$\leq \varepsilon^{-p} A \sum_{k=1}^{n} E||V_{nk}||^p + (**)^p$$

$$\leq \varepsilon^{-p} A \sum_{k=1}^{n} |a_{nk}|^p \int_0^{|a_{nk}|^{-1}} x^p dP[||V_k|| < x]$$

$$\leq \varepsilon^{-p} A \sum_{k=1}^{n} |a_{nk}|^p p \int_0^{|a_{nk}|^{-1}} x^{p-1} P[|V| \geq x] dx$$

$$= p\varepsilon^{-p} A \sum_{k=1}^{n} |a_{nk}|^r (|a_{nk}|^{p-r} \int_0^{|a_{nk}|^{-1}} x^{p-1} P[|V| \geq x] dx)$$

$$< p\varepsilon^{-p} A \Gamma(2\delta)$$

for n large enough so that $(\frac{1}{|a_{nk}|})^{r-p} \geq (\frac{1}{\max_k |a_{nk}|})^{r-p} > M(\delta)$. If the truncated means are not zero, then

$$(**)^p = ||\sum_{k=1}^{n} EV_{nk}|| = ||\sum_{k=1}^{n} E(a_{nk}V_k I_{[||a_{nk}V_k|| < 1]})||$$

$$= ||-\sum_{k=1}^{n} E(a_{nk}V_k I_{[||a_{nk}V_k|| \geq 1]})||$$

$$\leq \sum_{k=1}^{n} |a_{nk}| E||V_k I_{[||V_k|| \geq |a_{nk}|^{-1}]}||. \tag{5.3.15}$$

When $r = 1$, then (5.3.15) is less than or equal to

$$\sum_{k=1}^{n} |a_{nk}| E(|V| I_{[|V| \geq |a_{nk}|^{-1}]})$$

which goes to zero as $n \to \infty$. For $r > 1$, (5.3.15) is

$$\leq \sum_{k=1}^{n} |a_{nk}| \int_{|a_{nk}|^{-1}}^{\infty} P[|V| \geq x] dx \leq \sum_{k=1}^{n} |a_{nk}| \int_{|a_{nk}|^{-1}}^{\infty} \delta x^{-r} dx$$

$$\leq \frac{r}{r-1} \delta \sum_{k=1}^{n} |a_{nk}|^r$$

using (5.3.13) which also can be made small. ///

Note that in Theorem 5.3.2 the moment condition is significantly reduced to an rth moment with $0 < r < p$. With $p > 1$ and $r = 1$, this theorem implies a weak law of large numbers, but not a strong law. In Example 5.2.3 the strong law of large numbers did not hold for the random variables $\{V_n\}$ in R. However, $E|V_n| = 1/\log(n+2) \to 0$ as $n \to \infty$, and hence

$$E|\frac{1}{n} \sum_{k=1}^{n} V_k| \leq \frac{1}{n} \sum_{k=1}^{n} E|V_k| \to 0 \tag{5.3.15}$$

as $n \to \infty$. Since $EV_n = 0$ for all n, (5.3.15) implies that the weak law of large numbers hold. Without the geometry condition of type p, Theorem 5.3.2 is no longer valid as was illustrated by Example 5.2.2 for the space ℓ^1.

5.4 PROBLEMS

5.1 Let X be a finite-dimensional normed linear space and let $\{V_n\}$ be random elements in X such that

$$(*) \quad \sup_n E(||V_n||^r) < \infty \text{ for some } r > 0.$$

(a) Show that (*) implies that the random elements are tight.

(b) If $r \geq 1$, show that (*) implies that the random elements $\{V_n - EV_n\}$ are tight.

(c) If $r < 1$ in (*), show that the random elements $\{V_n - EV_n\}$ need not be tight even when EV_n exist for each n.

5.2 Let V be a symmetric random element in a separable normed linear space. Show that EV = 0 when it exists.

5.3 Show that $E|X_k| \leq E|X|$ for each k when

$$P[|X_k| \geq a] \leq P[|X| \geq a]$$

for each $a \geq 0$ and each k and when $E|X| < \infty$.

5.4 In the proof of Theorem 5.2.7 verify that $x_n \to 0$ in τ_S if and only if $||x_n|| \to 0$.

5.5 Let V be a random element in a separable Banach space such that $P[V \in K] = 1$ where K is a symmetric, compact, convex set with $0 \in K$. Show that $EV \in K$.

5.6 Show that convergence in probability for Theorem 5.3.1 follows as an easy corollary to Theorem 5.3.2.

CHAPTER VI

RANDOMLY WEIGHTED SUMS

6.0 INTRODUCTION

Because of the random nature of many problems in the applied sciences, researchers are increasingly switching from deterministic to probabilistic approaches. Thus, it is important to consider random weighting of random variables, that is, to allow $\{a_{nk}\}$ to be a double array of random variables. The results in this chapter pertain to these considerations. Without requiring any uncorrelation conditions between the random weights and random elements, convergence results with probability one and in the rth mean are obtained for randomly weighted sums.

6.1 IDENTICALLY DISTRIBUTED RESULTS

Theorem 6.1.2 [of Wei and Taylor (to appear, 1978 a)] will be an extension of Theorem 3.4.13 to random elements in normed linear spaces in addition to allowing the weights to be random variables. First, an inequality will be proved which is needed in the proofs of the main results.

Lemma 6.1.1 For positive numbers $x_1,\dots,x_n$ and $1 \le p \le q < \infty$,

$$(\textstyle\sum_{i=1}^{n} x_i^q)^{1/q} \le (\sum_{i=1}^{n} x_i^p)^{1/p}. \tag{6.1.1}$$

Proof: By induction, it must only be proved that

$$(x^q + y^q)^{1/q} \le (x^p + y^p)^{1/p}, \tag{6.1.2}$$

where x,y are positive. By a derivative test, it follows that

$$1 + t^r \le (1 + t)^r \tag{6.1.3}$$

for $r \ge 1$ and $t \ge 0$. The substitutions

$$t = y^p/x^p \text{ and } r = q/p \tag{6.1.4}$$

lead to the conclusion. ///

<u>Theorem 6.1.2</u> Let $\{V_n\}$ be independent identically distributed random elements in a separable normed linear space X such that $E||V_1||^r < \infty$ for some $r \in [1,2)$. Let $\{a_{nk}\}$ be random variables satisfying for some constant Γ,

$$P[\limsup_n \textstyle\sum_{k=1}^n a_{nk}^2 \le \Gamma] = 1 \tag{6.1.5}$$

with probability one for some constant Γ. Then

$$n^{-1/r} \textstyle\sum_{k=1}^n a_{nk} V_k \to 0 \tag{6.1.6}$$

with probability one.

<u>Proof</u>: Since

$$||n^{-1/r} \textstyle\sum_{k=1}^n a_{nk}V_k|| \le n^{-1/r} \textstyle\sum_{k=1}^n |a_{nk}| \; ||V_k||$$

and $\{||V_n||\}$ are random variables, $\{V_n\}$ can be assumed to be random variables without loss of any generality. Let $\varepsilon \in (0,1/2)$ be given. By the dominated convergence theorem, there exist $M > 0$ such that

$$E|V_1|^r I_{[|V_1| > M]} < (\varepsilon/2\Gamma^{1/2})^r. \tag{6.1.7}$$

For each k, let

$$U_k = V_k I_{[|V_k| \le M]} \text{ and } W_k = V_k I_{[|V_k| > M]}.$$

Also, let

$$S_n = n^{-1/r}\sum_{k=1}^{n} a_{nk}U_k,$$

and

$$T_n = n^{-1/r}\sum_{k=1}^{n} a_{nk}W_k.$$

Then, Inequality (6.1.7) becomes

$$E|W_1|^r < (\varepsilon/2\Gamma^{1/2})^r. \tag{6.1.8}$$

By the strong law of large numbers,

$$\frac{1}{n}\sum_{k=1}^{n}|W_k|^r \to E|W_1|^r \tag{6.1.9}$$

with probability one. Hence, the Cauchy-Schwartz Inequality and Lemma 6.1.1 imply that

$$\begin{aligned} |T_n|^r &= \frac{1}{n}(|\sum_{k=1}^{n} a_{nk}W_k|^2)^{r/2} \\ &\le \frac{1}{n}(\sum_{k=1}^{n} a_{nk}^2)^{r/2}(\sum_{k=1}^{n} W_k^2)^{r/2} \\ &\le (\sum_{k=1}^{n} a_{nk}^2)^{r/2}\frac{1}{n}\sum_{k=1}^{n}|W_k|^r. \end{aligned} \tag{6.1.10}$$

Hence, by (6.1.5), (6.1.8), (6.1.9) and (6.1.10),

$$\limsup_n |T_n|^r \le \Gamma^{r/2}\, E|W_1|^r \le (\varepsilon/2)^r \tag{6.1.11}$$

with probability one. Next, since $|U_k| \le M$,

$$\begin{aligned} |S_n|^r &\le (\sum_{k=1}^{n} a_{nk}^2)^{r/2} n^{-1}(\sum_{k=1}^{n} U_k{}^2)^{r/2} \\ &\le (\sum_{k=1}^{n} a_{nk}^2)^{r/2} n^{-(2-r)/2}(n^{-1}\sum_{k=1}^{n} U_k{}^2)^{r/2} \\ &\le (\sum_{k=1}^{n} a_{nk}^2)^{r/2} n^{-(2-r)/2} M^r. \end{aligned} \tag{6.1.12}$$

Since $r < 2$,

$$\limsup_n |S_n| = 0 \tag{6.1.13}$$

with probability one. A countable union of null sets can be

excluded in (6.1.11) and (6.1.13) by choosing a sequence $\varepsilon_m \to 0$ (and hence $M_m \to \infty$) to obtain a null set Ω_0. Thus, for $\omega \notin \Omega_0$ and $\delta > 0$, there exists $N(\delta,\omega)$ such that $n \geq N(\delta,\omega)$ implies that $|T_n(\omega)| < 2\delta/3$ and $|S_n(\omega)| < \delta/3$. ///

To obtain sharper convergence results of weighted sums when $\{a_{nk}\}$ are random weights instead of constants, the correlation of a_{nk} and V_k must be considered. The following simple example shows that the results of Chapter V cannot be extended directly to random weights.

Example 6.1.1 Let $\{V_k\}$ be independent, identically distributed random variables such that $V_1 = \pm 1$ each with probability 1/2. Let $a_{nk} = \frac{1}{n} V_k$ for $1 \leq k \leq n$, $a_{nk} = 0$ otherwise. Then $\{a_{nk}\}$ is a Toeplitz sequence of random variables such that $\sum_{k=1}^{\infty} |a_{nk}(\omega)| \leq 1$ and $\lim_{n\to\infty} a_{nk}(\omega) \to 0$ for each $\omega \in \Omega$. But, $|\sum_{k=1}^{\infty} a_{nk} V_k| \equiv 1 \not\to 0$. ///

It is important to observe that Condition (6.1.5) is not as restrictive as its appearance might indicate. For example, the weights (or in the order of)

$$a_{nk} = \begin{cases} \frac{1}{n} & \text{for } 1 \leq k \leq n \\ 0 & \text{for } k > n \end{cases}$$

not only satisfy Condition (6.1.5) but

$$\sum_{k=1}^{n} a_{nk}^2 = \frac{1}{n} \to 0 \qquad \text{as } n \to \infty.$$

The choice of weights (in the order of)

$$a_{nk} = \begin{cases} \frac{1}{\sqrt{n}} & \text{for } 1 \leq k \leq n \\ 0 & \text{for } k > n \end{cases}$$

also satisfy Condition (6.1.5) and illustrates the importance of $r < 2$.

6.2 CONVERGENCE IN THE rth MEAN FOR RANDOMLY WEIGHTED SUMS

The following theorem [from Wei and Taylor (to appear, 1978b)] concerns convergence in the rth mean, and hence a weak convergence may then be obtained as a corollary.

Theorem 6.2.1 Let $\{V_n\}$ be a random elements in a separable normed linear space X such that

$$\limsup_{n\to\infty} \frac{1}{n^r} E(\sum_{k=1}^n ||V_k||^2)^r = A < \infty \tag{6.2.1}$$

for some $1 \le r < 2$. Let $\{a_{nk}\}$ be random variables such that

$$\limsup_{n\to\infty} \sum_{k=1}^n (E|a_{nk}|^{2r})^{1/r} = B < \infty. \tag{6.2.2}$$

Then

$$\frac{1}{n^r} \sum_{k=1}^n a_{nk}V_k \to 0 \tag{6.2.3}$$

in the rth mean.

Proof: Since

$$\begin{aligned} E||n^{-1/r}\sum_{k=1}^n a_{nk}V_k||^r &\le \frac{1}{n} E[(\sum_{k=1}^n a_{nk}^2)^{r/2}(\sum_{k=1}^n ||V_k||^2)^{r/2}] \\ &\le \frac{1}{n} [E(\sum_{k=1}^n a_{nk}^2)^r \; E(\sum_{k=1}^n ||V_k||^2)^r]^{1/2} \\ &\le n^{-(2-r)/2}[(\sum_{k=1}^n (E|a_{nk}|^{2r})^{1/r})^r \\ &\quad \cdot E(\frac{1}{n} \sum_{k=1}^n ||V_k||^2)^r]^{1/2}, \end{aligned}$$

Conditions (6.2.1) and (6.2.2) imply (6.2.3). ///

The following example from regression analysis illustrates the results for randomly weighted sums.

<u>Example 6.2.1</u> In a general linear regression model

$$y_i = \alpha + \beta x_i + \varepsilon_i \tag{6.2.4}$$

where $\{\varepsilon_i\}$ are independent, identically distributed with mean 0 and variance σ^2, $i = 1,\ldots,n$. The usual least squares estimator gives

$$\hat{\beta} = \beta + \textstyle\sum_{i=1}^{n}[(x_i - \bar{x})/(\sum_{j=1}^{n}(x_j - \bar{x})^2]\varepsilon_i, \tag{6.2.5}$$

where $\bar{x} = \frac{1}{n}\sum_{i=1}^{n}x_i$. In many cases, x_i would be a random variable depending on $\{\varepsilon_i\}$. For example, in the clinical trials, x_i may be the dose level, and ε_i may be the resulting toxicity so that x_i depends on $\varepsilon_1,\ldots,\varepsilon_{i-1}$. Let

$$b_{ni} = (x_i - \bar{x})/\textstyle\sum_{i=1}^{n}(x_i - \bar{x})^2.$$

Then the convergence of $\sum_{i=1}^{n}b_{ni}\varepsilon_i$ to zero with probability one gives the strong consistency of the extimator $\hat{\beta}$. By Theorem 6.1.2 a sufficient condition for this convergence may be given as:

$$\limsup_{n\to\infty} n^r/\textstyle\sum_{i=1}^{n}(x_i - \bar{x})^2 \leq \Gamma \tag{6.2.6}$$

with probability one for some $r \in (1,2]$. ///

Note that this result holds without requiring any independent or uncorrelation assumptions between $\{x_i\}$ and $\{\varepsilon_i\}$. It is instructive to compare this result with Theorem 2 of Anderson and Taylor (1976). Using the independence of x_i with ε_i (but not necessarily with $\varepsilon_1,\ldots,\varepsilon_{i-1}$), they obtained the result when

$$\textstyle\sum_{i=1}^{n}(x_i - \bar{x})^2 \to \infty$$

with probability one.

Finally, these results can be extended to certain Fréchet spaces by analogous arguments to the one used in Chapter V.

6.3 PROBLEMS

6.1 Is Lemma 6.1.1 still valid when $0 < p \le q < \infty$?

6.2 In the proof of Theorem 6.2.1, justify the inequality

$$E(\sum_{k=1}^{n} a_{nk}^{2})^{r} \le [(\sum_{k=1}^{n}(E|a_{nk}|^{2r})^{1/r}]^{r}.$$

CHAPTER VII

LAWS OF LARGE NUMBERS IN D[0,1]

7.0 INTRODUCTION

In this chapter laws of large numbers for D[0,1] will be presented. Existing laws of large numbers for normed linear spaces can be used when the range of the random element is separable (with probability one) in the uniform topology on D[0,1]. The peculiar structure and properties of D[0,1] under the Skorohod topology have impeded development of the laws of large numbers. However, by introducing particular techniques laws of large numbers are obtained under hypotheses specifically involving the Skorohod topology.

In Section 7.1 the basic properties and notation for D[0,1] are presented. The geometric structure of the Skorohod topology is investigated, and the peculiar properties of the Skorohod metric are portrayed by several lemmas and examples. Also in Section 7.1 random elements in D[0,1], expected values, and Borel sets of the Skorohod and uniform topologies are considered.

In Section 7.2 weak laws of large numbers are obtained. The concept of convex tightness is introduced to obtain these results. For the spaces C[0,1] and C[S], the results of Section 7.2 provide more general results than the weak laws of large numbers in Sections 4.1 and 5.2.

In Section 7.3 strong laws of large numbers are proved. The role of compact sets and convex tightness is extremely important in obtaining these results. In addition, a "Chung-type" strong law of large numbers is proved for a class of random elements which include the sample distribution functions. Finally, Section 7.4 concludes with a detailed characterization of convex tightness.

7.1 PRELIMINARIES

Let $D = D[0,1]$ denote the space of real-valued functions on $[0,1]$ having discontinuities of the first kind, and hence let each function of D be right continuous and have left-hand limits at each point. For convenience, let $f(1) = \lim_{t\to 1^-} f(t)$.

Let Λ denote the space of one-to-one functions from $[0,1]$ onto $[0,1]$. The Skorohod metric d is defined by

$$d(x,y) = \inf \{\varepsilon > 0: \sup_{0\le t\le 1} |x(\lambda t) - y(t)| \le \varepsilon, \sup_{0\le t\le 1} |\lambda t - t| \le \varepsilon \text{ and } \lambda \in \Lambda\} \tag{7.1.1}$$

for $x,y \in D$. Detailed geometric and probabilistic properties of the space D with the Skorohod metric can be found in Billingsley (1968), pages 109-153 or Hahn (1975). However, a few properties basic to obtaining the laws of large numbers will be presented in this section.

The uniform norm

$$||x||_\infty = \sup_{0\le t\le 1} |x(t)| \tag{7.1.2}$$

can be defined on D but results in a non-separable Banach space. When relativized to the subspace $C[0,1]$ of D, the Skorohod topology coincides with the uniform topology on $C[0,1]$.

Since the Banach space obtained by giving D the uniform topology is not separable, the range of a sequence $\{X_n\}$ of random elements in D may not be separable in D with the uniform topology. Thus, the possibility of applying the many known results on laws of large numbers in Banach spaces is excluded by the nonseparability and since D with the Skorohod topology is not a Banach space nor even a Fréchet space. Providing D with the topology given by the Skorohod metric d makes the space D separable and eliminates this difficulty, but there are troublesome aspects to this structure. For example, D with the Skorohod metric is not complete, however, it is completely metrizable, that is, there is an equivalent metric d_o which makes D complete. The following properties provide more serious difficulties.

Lemma 7.1.1 $|\alpha| d(x,y) \leq d(\alpha x, \alpha y)$, for $|\alpha| \leq 1$, and $|\alpha| d(x,y) \geq d(\alpha x, \alpha y)$, for $|\alpha| \geq 1$.

Proof: $d(\alpha x, \alpha y) = \inf \{\varepsilon > 0: \sup_{0 \leq t \leq 1} |\alpha x(\lambda t) - \alpha y(t)| \leq \varepsilon$ and $\sup_{0 \leq t \leq 1} |\lambda t - t| \leq \varepsilon$, for some $\lambda \in \Lambda\}$

$= \inf \{\varepsilon > 0: \sup_{0 \leq t \leq 1} |x(\lambda t) - y(t)| \leq \frac{\varepsilon}{|\alpha|}$ and $\sup_{0 \leq t \leq 1} |\lambda t - t| \leq \varepsilon$, some $\lambda \in \Lambda\}$

$\geq \inf \{\varepsilon > 0: \sup_{0 \leq t \leq 1} |x(\lambda t) - y(t)| \leq \frac{\varepsilon}{|\alpha|}$ and $\sup_{0 \leq t \leq 1} |\lambda t - t| \leq \frac{\varepsilon}{|\alpha|}$, some $\lambda \in \Lambda\}$

$= |\alpha| \, d(x,y)$, and similarly for the second inequality. ///

An immediate and unfortunate (for the laws of large numbers since the reverse inequality is more desired) corollary is the following lemma.

Lemma 7.1.2 For any $x_i, y_i \in D$, $1 \leq i \leq n$,

$$\frac{1}{n} d(\textstyle\sum_{i=1}^n x_i, \sum_{i=1}^n y_i) \leq d(\frac{1}{n}\sum_{i=1}^n x_i, \frac{1}{n} \sum_{i=1}^n y_i).$$

Lemma 7.1.3 $d(0, \frac{1}{n} \sum_{i=1}^n x_i) \leq \frac{1}{n} \sum_{i=1}^n d(0,x_i)$, when $x_i \in D$, $1 \leq i \leq n$.

Proof: Follows from $d(0,\alpha x) = |\alpha| d(0,x)$ and $d(0,x+y) \le d(0,x) + d(0,y)$. ///

Lemma 7.1.4 If $x,y,u,v \in D$, then

$$d(x + u, y + v) \le d(x,y) + ||u||_\infty + ||v||_\infty .$$

Proof: Given $\varepsilon > 0$, find $\lambda \in \Lambda$ such that

$$\sup_{0\le t\le 1} |x(\lambda t) - y(t)| < d(x,y) + \varepsilon$$

and

$$\sup_{0\le t\le 1} |\lambda t - t| < d(x,y) + \varepsilon .$$

Thus,

$$|x(\lambda t) + u(\lambda t) - (y(t) + v(t)) + v(t) - u(\lambda t)| < d(x,y) + \varepsilon,$$

and

$$-|u(\lambda t) - v(t)| + |x(\lambda t) + u(\lambda t) - (y(t) + v(t))| < d(x,y) + \varepsilon .$$

Hence,

$$\sup_{0\le t\le 1} |x(\lambda t) + u(\lambda t) - (y(t) + v(t))| \le d(x,y) \sup_{0\le t\le 1} |u(\lambda t) - v(t)| + \varepsilon$$

$$\le d(x,y) + ||u||_\infty + ||v||_\infty + \varepsilon .$$

Recalling that $\sup_{0\le t\le 1} |\lambda t - t| < d(x,y) + \varepsilon$, it follows that

$$d(x + u, y + v) \le d(x,y) + ||u||_\infty + ||v||_\infty$$

since $\varepsilon > 0$ is arbitrary. ///

Remark: If $u = v = c$ = constant, then

$$d(x + c, y + c) = d(x,y).$$

With the Skorohod metric d, D is a linear space of functions and is a metric space, but it is not a linear topological space. The metric d is not translation invariant, and addition is not a continuous operation [Billingsley (1968), page 123]. However, scalar multiplication is continuous.

Example 7.1.1 Let $x_n = I_{[\frac{1}{2}-\frac{1}{n},1]}$, $n = 2,3,\ldots$, $x_0 = I_{[\frac{1}{2},1]}$, and $x = -x_0$. Then, $x_n \to x_0$ in the Skorohod topology, but $x_n + x = x_n - x_0 \not\to x_0 - x_0 = 0$ since

$$d(x_n + x, 0) = \sup_{0\le t\le 1} |I_{[\frac{1}{2}-\frac{1}{n},\frac{1}{2})}(t)| = 1$$

for all n. ///

However, the following interesting property holds.

Property 7.1.5 If $x_n \to x$, $y_n \to y$, and $x_n + y_n \to z$ with $x_n, y_n, x, y, z \in D$, then $z = x + y$.

Proof: For $x \in D$, let I_x denote the set of continuity points of x in $I = [0,1]$. For $t \in I_x \cap I_y \cap I_z$ we have by hypothesis $x_n(t) \to x(t)$, $y_n(t) \to y(t)$, and $x_n(t) + y_n(t) \to z(t)$ [Skorohod convergence implies convergence at continuity points] and thus $z(t) = x(t) + y(t)$. But the complement of $I_x \cap I_y \cap I_z$ is at most a countable subset of I and since $z \in D$ by hypothesis, $z = x + y$. ///

With the Skorohod topology, D is not locally convex. It is locally convex at some points, for example, at the origin and at the constant functions. Moreover, the neighborhood ball $B_\varepsilon(x_0) = \{x \in D: d(x,x_0) \le \varepsilon\}$ is not necessarily convex. In fact, the convex hull of $B_\varepsilon(x_0)$ may contain points x whose distance $d(x,x_0)$ from x_0 is arbitrarily large. Let $A \ge 1$ and let $0 < \varepsilon < \frac{1}{2}$.

Define $x_0 = AI_{[\frac{1}{2},1]}$, $x = AI_{[\frac{1}{2}-\varepsilon,1]}$, and $y = AI_{[\frac{1}{2}+\varepsilon,1]}$. Note that $d(x_0,x) = \varepsilon = d(x_0,y)$, but $d(x_0, \frac{x+y}{2}) \geq \frac{1}{2}A$ since $\sup_{0 \leq t \leq 1} |x_0(\lambda t) - (\frac{x+y}{2})(t)| \geq \frac{1}{2}A$ for any $\lambda \in \Lambda$. Also, A can be arbitrarily large.

Recall the characterization of compact sets in D which is given by Billingsley (1968), page 116. A set $K \subset D$ is relatively compact if and only if

$$\text{(i)} \quad \sup_{x \in K} ||x||_\infty < \infty$$

$$\text{(ii)} \quad \lim_{\delta \to 0} \sup_{x \in K} w'_x(\delta) = 0 \tag{7.1.3}$$

where $w'_x(\delta)$ is the modulus of continuity defined by

$$w'_x(\delta) = \inf_{\{t_j\}} \max_{1 \leq j \leq r} \sup_{s,t \in [t_{j-1},t_j)} |x(t) - x(s)| \tag{7.1.4}$$

where $0 = t_0 < t_1 < \ldots < t_r = 1$ with $t_j - t_{j-1} > \delta$.

<u>Example 7.1.2</u> The convex hull of even the simplest compact set in D need not be relatively compact. Let $x_s = I_{[s,1]}$, for $0 \leq s < 1$, and define the set

$$K = \{x_s: \ 0 < a \leq s \leq 1-a < 1\}, \ 0 < a < \tfrac{1}{2}. \tag{7.1.5}$$

Note, K is relatively compact in D: (i) is obviously satisfied and $w'_x(\delta) = 0$ whenever $0 < \delta < a$ for any $x \in K$, so that $\lim_{\delta \downarrow 0} \sup_{x \in K} w'_x(\delta) = 0$, and (ii) is satisfied.

Let $\frac{1}{2} < A < 1$. For any δ, $0 < \delta < \frac{1}{2}$, define

$$x_\delta = Ax_{\frac{1}{2}} + (1-A)x_{\frac{1}{2}+\delta}.$$

For any δ, $0 < \delta < \frac{1}{2} - A$, $x_\delta \in K_1$, the convex hull of K, and

$w'_{x_\delta}(\delta) \geq A$, and hence $\sup_{x \in K_1} w'_x(\delta) \geq w'_x(\delta) \geq A$. Hence (ii) is not true for K_1, and K_1 is not relatively compact. ///

Example 7.1.2 can be strengthened by making the set K countably infinite.

A random element in D equipped with the Skorohod topology is a measurable map X: $\Omega \to D$ from a probability space (Ω, F, P) into D. The Borel sets of the Skorohod topology in D are denoted by $\mathcal{D}$, and the Borel sets of the uniform topology are denoted by $\mathcal{U}$. Then X: $\Omega \to D$ is a random element in D if $\{\omega\colon X(\omega) \in B\} \in F$ for every $B \in \mathcal{D}$. Random elements in D are characterized as follows: X is a random element in D if and only if X(t) is a random variable for each $t \in [0,1]$ [Billingsley (1968), page 128].

Let X be a random element in D and let

$$E[\sup_{0 \leq t \leq 1} |X(t)|] = E||X||_\infty < \infty \tag{7.1.6}$$

[$||X||_\infty$ is a random variable since $||X||_\infty = \sup_{t \in Q} |X(t)|$ where Q are the rationals in [0,1].] For a random element X which satisfies (7.1.6), the expected value can be defined pointwise by

$$(EX)(t) = E[X(t)] \quad \text{for each } t \in [0,1].$$

For each $t \in [0,1]$, $|X(t)| \leq ||X||_\infty$, and (EX)(t) exists. By the Lebesgue dominated convergence theorem,

$$\lim_{s \to t^-} E[X(s)] = E[X(t - 0)]$$

and

$$\lim_{s \to t^+} E[X(s)] = E[X(t)].$$

Thus, $EX \in D$ when $E||X||_\infty < \infty$.

Example 7.1.3 The following example shows that the condition $E||X|| < \infty$ is not necessary for $EX \in D$. Put $x_n = 2^n I_{E_n}$, where $E_n = [\frac{1}{2} + \frac{1}{2^{n+1}}, \frac{1}{2} + \frac{1}{2^n})$, and define X by $P[X = x_n] = 2^{-n-1} = P[X = -x_n]$. Clearly, $EX(t) = 0$ for each $t \in [0,1]$, and so $EX \in D[0,1]$. But, $E[\sup_{0 \le t \le 1} |X(t)|] = \sum_{n=1}^{\infty} 2^n \cdot 2^{-n} = +\infty$. ///

Having seen that the condition $E||X||_\infty < \infty$ is not a necessary condition for EX [as defined by $(EX)(t) = E[X(t)]$] to be an element of D, it is natural to ask to what extent this condition can be weakened and still yield $EX \in D$. That the condition $E||X||_\infty < \infty$ is very close to being necessary and is probably the weakest "simple" condition giving $EX \in D$ is illustrated by the following example.

Example 7.1.4 The following example shows that the condition $\sup_{0 \le t \le 1} E|X(t)| < \infty$ does not imply that $EX \in D[0,1]$. Let $x_n = (-1)^n 2^n I_{E_n}$ where $E_n = [\frac{1}{2} + \frac{1}{2^{n+1}}, \frac{1}{2} + \frac{1}{2^n})$, for $n = 2,3,\ldots$, and $I_1 = [\frac{3}{4}, 1]$. Define the random element X by $P[X = x_n] = 2^{-n}$. Then

$$E|X(t)| = \begin{cases} 2^n \cdot 2^{-n} = 1 & \text{if } t \ge \frac{1}{2} \\ 0 & \text{if } t < \frac{1}{2}, \end{cases}$$

and so $\sup_{0 \le t \le 1} E|X(t)| = 1$.

But,

$$EX(t) = \begin{cases} -1 & \text{if } t \in E_{2n+1} \\ +1 & \text{if } t \in E_{2n}\,. \end{cases}$$

Thus, $\overline{\lim}_{t \to \frac{1}{2}^+} EX(t) = 1 > -1 = \underline{\lim}_{t \to \frac{1}{2}^+} EX(t)$, and so EX cannot be in $D[0,1]$.

Of course, $E[\sup_{0 \le t \le 1} |X(t)|] = \sum_{n=1}^{\infty} 2^n \cdot 2^{-n} = +\infty$, as it must be.

Note that $E|X(t)|$ is bounded on $[0,1]$ in this example. ///

Consider the set $\{x_s = I_{[s,1]}: \ 0 \le s < 1\} \subset D[0,1]$. Let $N_{\frac{1}{2}}(x_s) = \{y: \sup_{0 \le t \le 1} |y(t) - x_s(t)| < \frac{1}{2}\}$. Then $N_{\frac{1}{2}}(x_s) \cap N_{\frac{1}{2}}(x_t) = \emptyset$ for $s \ne t$. Let C denote the set of all unions of neighborhoods $N_{\frac{1}{2}}(x_t)$. Then, $\text{card}\ (C) = 2^{\text{card}\{N_{\frac{1}{2}}(x_s): \ 0 \le s < 1\}} = 2^c$. Note that C consists of open subsets and hence is contained in $\mathcal{U}$. Since $\text{card}\ (\mathcal{D})$ = the continuum, it follows that $\text{card}\ (\mathcal{D}) \underset{\ne}{<} \text{card}\ (\mathcal{U})$ and $\mathcal{D} \underset{\ne}{\subset} \mathcal{U}$.

Since both topologies are Hausdorff, a countably-valued random element is measurable with respect to both collection of Borel sets, $\mathcal{U}$ and $\mathcal{D}$. The following example provides a random element which is not measurable with respect to $\mathcal{U}$.

<u>Example 7.1.5</u> Let $x_s(t) = I_{[s,1]}(t)$ for $0 \le s < 1$, and let $S = \{x_s: \ 0 \le s < 1\}$. On S define a probability measure as follows:

$$P\{x_s: \ s_1 \le s < s_2\} = s_2 - s_1 \text{ for any } 0 \le s_1 < s_2 \le 1.$$

Then for any (Skorohod) Borel set B in D, $\{s \in [0,1]: \ x_s \in B\}$ is Borel measurable and by definition, $P(B) = m\{s: \ x_s \in B\}$.

To see that $\{s: \ x_s \in B\}$ is measurable, let U be any open set in D. Let $N_\varepsilon(x) = \{y \in D: \ d(x,y) < \varepsilon\}$. If $x_{s_o} \in U$, then $N_\varepsilon(x_{s_o}) \subset U$ for some $\varepsilon > 0$, and it is easy to see that $x_s \in N_\varepsilon(x_{s_o})$ whenever $|s - s_o| < \varepsilon$ [this would not be true of the uniform norm]. Hence, $\{s: \ x_s \in N_\varepsilon(x_{s_o})\}$ is connected and is an open interval in I, hence Borel measurable. Hence, $\{s: \ x_s \in U\}$ is a union of open intervals and is Borel measurable.

Now let B be a Borel set in D. $\{s: \ x_s \in B\}$ is generated in the same way from open sets in $[0,1]$ as B is generated by open sets in D. Hence, $\{s: \ x_s \in B\}$ is a Borel set on $[0,1]$.

Let E be a non-measurable subset of [0,1]. Letting $\tilde{N}_\varepsilon(x) = \{y: ||x - y||_\infty < \varepsilon\}$, we have that the set

$$S = \bigcup_{s \in E} \tilde{N}_\varepsilon(x_s)$$

is open in D. Since, for $0 < \varepsilon < 1$, x_s is the only element of S which is in $\tilde{N}_\varepsilon(x_s)$, $s \neq s'$ implies that $\tilde{N}_\varepsilon(x_s) \cap \tilde{N}_\varepsilon(x_{s'}) = \emptyset$. But then the probability of the set $\tilde{S}$ is not defined, since its probability would be the measure of the set $\{s \in [0,1]: x_s \in \tilde{S}\} = E$.

Thus, defining a random element X in D as above, using the uniform distribution on [0,1] (which can then be taken for the probability space Ω), X will be a random element in D with the Skorohod topology, but not in D with the uniform topology. The mapping X into D with the uniform topology is neither separably valued nor Borel measurable. ///

For each positive integer m, let $E_{i,m} = [\frac{i-1}{2^m}, \frac{i}{2^m})$ for $i = 1,2,\ldots,2^m-1$ and let $E_{2^m,m} = [\frac{2^m-1}{2^m}, 1]$. Define the operator T_m: $D \to D$ by

$$T_m(x) = \sum_{i=1}^{2^m} x(\tfrac{i-1}{2^m}) I_{E_{i,m}} . \qquad (7.1.7)$$

Each T_m is additive: $T_m(ax + by) = aT_m(x) + bT_m(y)$. Moreover, each T_m is a projection of D onto a closed linear subspace D_m, and D is the closure of $\bigcup_{m=1}^{\infty} D_m$ in the Skorohod topology [the closure of $\bigcup_{m=1}^{\infty} D_m$ in the uniform topology is much smaller].

<u>Lemma 7.1.6</u> For each $x \in D$, $\lim_{m\to\infty} d(x,T_m(x)) = 0$.

<u>Proof</u>: Step functions of the form $x_m = \sum_{i=1}^{2^m} r_{i-1} I_{E_{i,m}}$ ($E_{i,m}$ as above) with r_i rational are dense in D with the Skorohod topology (Billingsley, page 112). Let $x \in D$. To every $\varepsilon > 0$ there is such an x_m and a $\lambda \in \Lambda$ satisfying $\sup_{0 \le t \le 1} |x(\lambda t) - x_m(t)| \le \varepsilon$

and $\sup_{0\le t\le 1} |\lambda t - t| \le \varepsilon$. Find r_i, $1 \le i \le 2^m$, such that

$\sup_{1\le i\le 2^m} |x(\frac{i}{2^m}) - r_i| \le \varepsilon$; thus $\sup_{t\in I} |T_m(x)(t) - x_m(t)| \le \varepsilon$. Then,

$$\begin{aligned}\sup_{t\in I} |T_m(x)(t) - x(t)| &\le \sup_{t\in I} |T_m(x)(t) - x_m(t)| \\ &\quad + \sup_{t\in I} |x_m(t) - x(\lambda t)| \\ &\le \varepsilon + \varepsilon = 2\varepsilon .\end{aligned}$$

Since $\varepsilon > 0$ is arbitrary, $\lim_{m\to\infty} d(x, T_m(x)) = 0$. ///

<u>Lemma 7.1.7</u> For $x \in D$, $d(x, T_m x) < w'_x(2^{-m}) + 2^{-m}$.

<u>Proof</u>: Given $x \in D$, there exists a partition $\{t_0, t_1, \ldots, t_{r_m}\}$ of $[0,1]$, $0 = t_0 < t_1 < \ldots < t_{r_m} = 1$, such that $2^{-m} < t_j - t_{j-1}$ and

$$\sup_{t_j\le s,t\le t_j} |x(t) - x(s)| < w'_x(2^{-m}) + 2^{-m},$$

for all $j = 0,1,\ldots,t_{r_m}-1$. For each t_j, let i_j be such that $\frac{i_{j-1}}{2^m} < t_j \le \frac{i_j}{2^m}$, for $j = 1,\ldots,t_{r_m}-1$. The i_j's are distinct and there exists $\lambda \in \Lambda$ that maps $\frac{i_j}{2^m}$ to t_j and $\lambda t \le t$ for every $t \in [0,1]$, and $\sup|\lambda t - t| < 2^{-m}$.

Next, $[\lambda(\frac{i}{2^m}), \lambda(\frac{i+1}{2^m}))$ is contained in some $[t_j, t_{j+1})$ for each $i = 0,1,\ldots,2^m-1$. Thus, for

$$T_m x = \sum_{i=1}^{2^m} x(\tfrac{i}{2^m}) I_{[\frac{i}{2^m},\frac{i+1}{2^m})},$$

it follows that

$$d(x,T_m x) \le \max\{2^{-m}, \sup_{0\le t\le 1} |x(t) - (T_m x)(\lambda^{-1}t)|\}.$$

But,

$$\sup_{0\le t\le 1} |x(t) - (T_m x)(\lambda^{-1}(t))|$$

$$= \sup_{0\le t\le 1} |x(t) - \textstyle\sum_{i=1}^{2^m} x(\frac{i-1}{2^m}) I_{[\frac{i-1}{2^m},\frac{i}{2^m})}(\lambda^{-1}t)|$$

$$= \sup_{0\le t\le 1} |x(t) - \textstyle\sum_{i=1}^{2^m} x(\frac{i-1}{2^m}) I_{[\lambda(\frac{i-1}{2^m}),\lambda(\frac{i}{2^m}))}(t)|$$

$$= \max_{1\le i\le 2^m} \sup_{t\in[\lambda(\frac{i-1}{2^m}),(\frac{i}{2^m}))} |x(t) - x(\tfrac{i-1}{2^m})|$$

$$\le \max_{1\le j\le r_m} \sup_{s,t\in[t_{j-1},t_j)} |x(t) - x(s)|$$

$$< w_x'(2^{-m}) + 2^{-m}, \text{ since for each } i$$

$[\lambda(\frac{i-1}{2^m}),\lambda(\frac{i}{2^m})) \subset [t_{j-1},t_j)$ for some j, and $t_{j-1} \le \lambda(\frac{i-1}{2^m}) \le \frac{i-1}{2^m}$

$< \lambda(\frac{i}{2^m}) \le t_j$. Since $\sup_t|\lambda t - t| < 2^{-m}$,

$$d(x,T_m x) < w_x'(2^{-m}) + 2^{-m}. \qquad ///$$

Two valuable results follow immediately as corollaries to Lemma 7.1.7.

<u>Corollary 7.1.8</u> If $K \subset D$ is compact, then

$$\lim_{n\to\infty} \sup_{x\in K} d(x,T_m x) = 0.$$

<u>Proof</u>: By Lemma 7.1.7

$$\sup_{x\in K} d(x,T_m x) < \sup_{x\in K} w_x'(2^{-m}) + 2^{-m}.$$

Now use (ii) of (7.1.3). ///

<u>Corollary 7.1.9</u> $d(T_m x,T_m y) < w_x'(2^{-m}) + w_y'(2^{-m}) + d(x,y) + 2^{-m+1}$

Proof: It follows easily from previous results and

$$d(T_m x, T_m y) \le d(T_m x, x) + d(x,y) + d(y, T_m y). \qquad ///$$

The last property to be presented in this section of preliminaries is concerned with the operation of the Borel function T_m on a random element X in D.

Lemma 7.1.10 Let X be a random element in D[0,1]. Then,

(i) $|T_m X(t)| \le T_m|X(t)|$, for each $t \in [0,1]$.

(ii) $E[T_m X] = T_m(EX)$, if EX exists.

Proof: (i) Let $x \in D$. If $x \ge 0$ then $T_m x \ge 0$, and so $|T_m x| = T_m x = T_m|x|$. Thus,

$$\begin{aligned} |T_m x| &= |T_m(x_+ - x_-)| = |T_m x_+ - T_m x_-| \\ &\le |T_m x_+| + |T_m x_-| \\ &= T_m x_+ + T_m x_- = T_m(x_+ + x_-) = T_m|x|. \end{aligned}$$

(ii) Let X be countably-valued with range $\{x_1, x_2, \ldots\}$ and put $p_i = P[X = x_i]$. Then, $X = \sum_{i=1}^{\infty} x_i I_{[X=x_i]}$ and $T_m X = \sum_{i=1}^{\infty} T_m x_i I_{[X=x_i]}$. Thus,

$$\begin{aligned} T_m(EX) &= T_m(\textstyle\sum_{i=1}^{\infty} x_i p_i) \\ &= \textstyle\sum_{i=1}^{\infty} T_m(x_i)p_i, \quad \text{by the existence of EX} \\ &= E[T_m X]. \end{aligned}$$

Hence, (ii) now follows by approximation. ///

7.2 WEAK LAWS OF LARGE NUMBERS FOR IN D[0,1]

Some weak laws of large numbers for D[0,1] are presented in this section. In particular, a version of Theorem 5.2.6 is obtained using the concept of convex tightness. Convex tightness coincides with the definition of tightness (see Definition 5.2.1) for a Banach space, and the results of this section have surprising ramifications when restricted to the subspace C[0,1] [or applied to C[S]] since pointwise conditions are sufficient for the weak law of large numbers.

Since D[0,1] is a metric space, the definitions of identical distributions and independence and the basic probabilistic properties are valid. However, the absence of translation invariance and local convexity necessitates the following definition of convex tightness to obtain laws of large numbers.

<u>Definition 7.2.1</u> A sequence of random elements $\{X_k\}$ in D[0,1] is said to be <u>convex tight</u> if for each $\varepsilon > 0$ there exists a convex, compact set K_ε such that $P[X_k \in K_\varepsilon] > 1 - \varepsilon$ for all k.

Without loss of generality, it can be assumed that 0 is in the convex, compact set of Definition 7.2.1 (see Problem 7.1).

<u>Theorem 7.2.1</u> [Taylor and Daffer (to appear)] Let $\{X_k\}$ be a sequence of convex tight random elements in D[0,1] such that

$$E(||X_k||_\infty^r) \le \Gamma \tag{7.2.1}$$

where $r > 1$ and $\Gamma > 0$. If

$$\frac{1}{n}\sum_{k=1}^{n} (X_k(t) - EX_k(t)) \to 0 \tag{7.2.2}$$

in probability for each dyadic rational in [0,1], then

$$d(\frac{1}{n}\sum_{k=1}^{n} X_k, \frac{1}{n}\sum_{k=1}^{n} EX_k) \to 0$$

in probability.

Proof: Let $0 < \varepsilon < 1$ and $0 < \delta < 1$ be given. Pick K convex, compact such that $0 \in K$ and

$$P[X_k \notin K] < (\varepsilon\delta/24)^{\frac{r}{r-1}} / \Gamma^{\frac{1}{r-1}} \quad (7.2.3)$$

for all k. Thus,

$$E||X_k I_{[X_k \notin K]}||_\infty \le (E||X_k||^r)^{\frac{1}{r}} P[X_k \notin K]^{\frac{r-1}{r}} \le \varepsilon\delta/24. \quad (7.2.4)$$

Next, pick m such that $d(x, T_m(x)) < \varepsilon/6$ for all $x \in K$. For each n

$$P[d(\frac{1}{n}\sum_{k=1}^{n} X_k, \frac{1}{n}\sum_{k=1}^{n} EX_k) > \varepsilon]$$

$$\le P[d(\frac{1}{n}\sum_{k=1}^{n} X_k, \frac{1}{n}\sum_{k=1}^{n} T_m(X_k)) > \varepsilon/3]$$

$$+ P[d(\frac{1}{n}\sum_{k=1}^{n} T_m(X_k), \frac{1}{n}\sum_{k=1}^{n} T_m(EX_k)) > \varepsilon/3]$$

$$+ P[d(\frac{1}{n}\sum_{k=1}^{n} T_m(EX_k), \frac{1}{n}\sum_{k=1}^{n} EX_k) > \varepsilon/3]. \quad (7.2.5)$$

Using Lemma 7.1.4 and (7.2.4), the first term of (7.2.5) can be expressed as

$$P[d(\frac{1}{n}\sum_{k=1}^{n} X_k, \frac{1}{n}\sum_{k=1}^{n} T_m(X_k)) > \varepsilon/3]$$

$$\le P[||\frac{1}{n}\sum_{k=1}^{n} X_k I_{[X_k \notin K]}||_\infty + ||\frac{1}{n}\sum_{k=1}^{n} X_k I_{[X_k \notin K]}||_\infty > \varepsilon/6]$$

$$\le (12/\varepsilon n)\sum_{k=1}^{n} E||X_k I_{[X_k \notin K]}||_\infty \le (12/\varepsilon)(\varepsilon\delta/24) = \delta/2. \quad (7.2.6)$$

Since K is convex, compact, and $0 \in K$, then

$$E(X_k I_{[X_k \in K]}) \in K \text{ and } \frac{1}{n}\sum_{k=1}^{n} E(X_k I_{[X_k \in K]}) \in K.$$

Moreover, $d(\frac{1}{n}\sum_{k=1}^{n} E(X_k I_{[X_k \in K]}), T_m(\frac{1}{n}\sum_{k=1}^{n} E(X_k I_{[X_k \in K]}))) < \varepsilon/6$ (7.2.7)

for each n. Thus, the third term of (7.2.5) can be expressed as

$$P[d(\tfrac{1}{n}\textstyle\sum_{k=1}^{n}T_m(EX_k),\ \tfrac{1}{n}\sum_{k=1}^{n}EX_k) > \varepsilon/3]$$

$$\le P[\|\tfrac{1}{n}\textstyle\sum_{k=1}^{n}T_m(EX_kI_{[X_k\notin K]}\|_\infty + \|\tfrac{1}{n}\sum_{k=1}^{n}EX_kI_{[X_k\notin K]}\|_\infty > \varepsilon/6].$$

But, $\|T_m x\|_\infty \le \|x\|_\infty$, and

$$\|\tfrac{1}{n}\textstyle\sum_{k=1}^{n}EX_kI_{[X_k\notin K]}\|_\infty \le \tfrac{1}{n}\sum_{k=1}^{n}\|EX_kI_{[X_k\notin K]}\|_\infty$$

$$\le \tfrac{1}{n}\textstyle\sum_{k=1}^{n}E\|X_kI_{[X_k\notin K]}\|_\infty < \varepsilon\delta/24.$$

Thus, the third term of (7.2.5) has probability zero. For the second term of (7.2.5),

$$P[d(\tfrac{1}{n}\textstyle\sum_{k=1}^{n}T_m(X_k),\ \tfrac{1}{n}\sum_{k=1}^{n}T_m(EX_k) > \varepsilon/3]$$

$$\le P[\|\tfrac{1}{n}\textstyle\sum_{k=1}^{n}(T_m(X_k) - T_m(EX_k))\|_\infty > \varepsilon/3]$$

$$\le \textstyle\sum_{i=1}^{2^m}P[|\tfrac{1}{n}\sum_{k=1}^{n}(X_k(\tfrac{i-1}{2^m}) - EX_k(\tfrac{i-1}{2^m}))| > \varepsilon/3.2^m]$$

$$< \delta/2 \text{ for } n \ge N \text{ from } (7.2.2). \tag{7.2.8}$$

From (7.2.5), (7.2.6), and (7.2.8) it follows that

$$P[d(\tfrac{1}{n}\textstyle\sum_{k=1}^{n}X_k,\ \tfrac{1}{n}\sum_{k=1}^{n}EX_k) > \varepsilon] < \delta$$

for all $n \ge N$. ///

Remark: If $\frac{1}{n}\sum_{k=1}^{n}EX_k$ converges to a constant, then Theorem 7.2.1 is if and only if since convergence in probability implies weak convergence (in distribution) and hence pointwise convergence in distribution to a constant which yields pointwise convergence in probability.

The bounded r^{th} $(r > 1)$ moments condition can not be reduced to a bounded first moments condition even for tight (hence convex tight) random elements in Banach spaces by Example 5.2.2. However, by requiring identical distributions, a first moment weak law of large numbers can be obtained in a similar proof to Theorem 7.2.1. Since convex compact sets, K_n, can be chosen so that $P[X_1 \in K_n] > 1 - \frac{1}{n^2}$, $X_1 I_{[X_1 \notin K_n]} \to 0$ with probability one. Thus, corresponding to (7.2.4) there exists a convex compact K such that $0 \in K$ and

$$E||X_k I_{[X_k \notin K]}||_\infty = E||X_1 I_{[X_1 \notin K]}|| < \varepsilon\delta/24$$

for all k. Hence, the proof of the following theorem is mutatis mutandis as the proof of Theorem 7.2.1.

<u>Theorem 7.2.2</u> Let $\{X_k\}$ be a sequence of identically distributed, convex tight random elements in D[0,1] such that $E||X_1|| < \infty$ For each dyadic rational $t \in [0,1]$ $\frac{1}{n}\sum_{k=1}^n X_k(t) \to EX_1(t)$ in probability if and only if $d(\frac{1}{n}\sum_{k=1}^n X_k, EX_1) \to 0$ in probability.

If EX_1 is continuous (for example, the mean function for a Poisson process on [0,1]), then

$$||\tfrac{1}{n}\textstyle\sum_{k=1}^n X_k - EX_1||_\infty = \sup_t |\tfrac{1}{n}\textstyle\sum_{k=1}^n X_k(t) - EX_1(t)| \to 0$$

in probability.

Before proceeding to the discussion of these results for the Banach spaces C[0,1] and C[S], the following corollary illustrates the pointwise condition by providing a sufficient condition for the hypotheses of Theorem 7.2.1.

<u>Corollary</u> Let $\{X_k\}$ be a sequence of convex tight random elements in D such that $E||X_k||_\infty^r \le \Gamma$ for all k, where $r > 1$ and Γ is a constant. If

(i) $Cov(X_\ell(t), X_k(t)) = 0$ for each $k \ne \ell$ and each $t \in [0,1]$

and

(ii) $\sum_{k=1}^n Var(X_k(t)) = o(n^2)$ for each t,

then

$$d(\frac{1}{n}\sum_{k=1}^{n}X_k,\ n^{-1}\sum_{k=1}^{n}EX_k) \to 0$$

in probability.

Proof: For each t, $\{X_k(t)\}$ is a sequence of uncorrelated random variables by (i). Next, (ii) is sufficient for the weak law of large numbers to hold for the random variables $\{X_k(t)\}$. ///

If the random element X has $\|\ \|_\infty$ - complete, separable support, then X is convex tight since P_X is basically a probability measure on a separable Banach space. However, Example 7.1.5 shows that not every random element on D[0,1] is convex tight. But, in the next section a strong law of large numbers (with convergence in the $\|\ \|_\infty$ topology) is obtained for independent, identically distributed random elements with the probability structure of Example 7.1.5 by a proof similar to that of the Glivenko-Cantelli Theorem.

On the subspace $C[0,1] \subset D[0,1]$, the Skorohod metric and the uniform norm are equivalent. Moreover, each probability measure on a complete, separable, metric space is tight [Billingsley (1968), page 10], and the concepts of convex tight and tight coincide for Banach spaces since the convex hull of a compact set is again compact. Thus, Theorems 7.2.1 and 7.2.2 provide more applicable results for C[0,1] than the earlier results of Taylor (1972) and Theorem 5.2.6 since only pointwise conditions must be verified (such as pointwise uncorrelation or pointwise independence).

Let C[S] denote the space of continuous, real-valued functions with domain S, a compact metric space. Let $\|x(s)\| = \sup_{s\in S}|x(s)|$ be the sup norm on C[S]. In obtaining weak laws of large numbers for C[0,1] of the form of Theorems 7.2.1 and 7.2.2, the linear, Borel approximating function T_m is the key development. Let

$\{s_1, s_2, s_3, \ldots\}$ denote a countable dense subset of S, and let $N(s_i, \varepsilon) = \{s \in S\colon\ d(s_i, s) < \varepsilon\}$. For each positive integer m there exists t_m such that

$$S = \bigcup_{i=1}^{t_m} N(s_i, \tfrac{1}{m}). \tag{7.2.9}$$

Let $\{f_i\colon\ i = 1, \ldots, t_m\} \subset C[S]$ be a locally finite partition of unity on S subordinate to the open covering in (7.2.9) [see Willard (1968), page 152]. For each m define

$$T_m(x) = \textstyle\sum_{i=1}^{t_m} x(s_i) f_i. \tag{7.2.10}$$

It then follows that each T_m is linear and continuous and $||T_m(x) - x|| \to 0$ for each $x \in C[S]$. Moreover, since $\sum_{i=1}^{t_m} f_i(s) = 1$ for each $s \in S$,

$$||T_m(x)|| \le ||x|| \quad \text{for each } x \in C[S].$$

Thus, $||T_m(x) - x|| \to 0$ uniformly for x in a compact set, and versions of Theorems 7.2.1 and 7.2.2 (assuming only tightness or identical distributions) follow for C[S].

7.3 STRONG LAWS OF LARGE NUMBERS IN D[0,1]

The strong law of large numbers by R. Ranga Rao (1963) for independent, identically distributed random elements will be presented first in this section. The next two strong laws of large numbers will be from Daffer and Taylor (1977). A version of Corollary 5.2.9 will be obtained for independent, convex tight random elements in D[0,1]. Finally, a strong law for independent random elements which take their values in the class of increasing functions is presented.

The following lemma plays a key role in proving the strong law of large numbers for independent, identically distributed random elements.

Lemma 7.3.1 Let X be a random element in D with $E||X|| < \infty$. For each $\varepsilon > 0$ there exists a partition $0 = t_0 < t_1 < \ldots < t_m = 1$ of [0,1] such that

$$\max_{1 \le i \le m} \sup_{s,t \in J_i} E|X(s) - X(t)| \le \varepsilon \tag{7.3.1}$$

where $J_i = [t_{i-1}, t_i)$.

Proof: For $0 \le \alpha < \beta \le 1$, let

$$\rho(\alpha,\beta) = \sup_{\alpha \le t,s < \beta} E|X(t) - X(s)|. \tag{7.3.2}$$

Let $\tau_1 = 1$ if $\rho(0,1) \le \varepsilon$; otherwise, let $\tau_1 = \inf\{t: \rho(0,t) > \varepsilon\}$. In general, let $\tau_j = 1$ if $\rho(\tau_{j-1},1) \le \varepsilon$; otherwise, let $\tau_j = \inf\{t: t > \tau_{j-1}$ and $\rho(\tau_{j-1},t) > \varepsilon\}$. Suppose (for a contradiction) that $\tau_j < 1$ for all j. Then there exists a sequence of points $\{t_n\}$ such that $\tau_n \le t_n < \tau_{n+1}$ and for each n $E|X(t_n) - X(\tau_{n+1})| > \varepsilon/2$. Since $\{\tau_n\}$ is a monotone increasing sequence which is bounded above by 1, $X(t_n) - X(\tau_{n+1}) \to 0$ pointwise in D[0,1]. Thus, by the dominated convergence theorem, $E|X(t_n) - X(\tau_{n+1})| > \varepsilon/2$ for each n is impossible. ///

Theorem 7.3.2 If $\{X_n\}$ are independent, identically distributed random elements in D[0,1] such that $E||X_1|| < \infty$, then

$$||\tfrac{1}{n} \textstyle\sum_{k=1}^{n} X_k - EX_1|| \to 0$$

with probability one.

In the Ranga Rao's proof of Theorem 7.3.2, the following characterization of compact sets in terms of $w_x''(\delta)$ [see Billingsley

(1968)] is used: if K is a compact set in D, then for $\varepsilon > 0$ there exists $\delta > 0$ such that $|x(t) - x(\alpha)| \le |x(\beta-0) - x(\alpha)| + \varepsilon$ whenever $\alpha \le t < \beta < \alpha + \delta$ and $x \in K$.

Proof of Theorem 7.3.2: Let $\varepsilon > 0$ be given. Since D is separable and complete, then P_{X_1} is tight and there exists a compact set K such that

$$E(\|X_1\|_\infty I_{[X_1 \notin K]}) \le \varepsilon. \tag{7.3.3}$$

Choose $\delta > 0$ so that $\alpha \le t < \beta < \alpha + \delta$ implies that

$$|x(t) - x(\alpha)| \le |x(\beta-0) - x(\alpha)| + \varepsilon \tag{7.3.4}$$

uniformly for $x \in K$. By Lemma 7.3.1 there exists a partition $\{t_i: i = 1,\dots,m\}$ of [0,1] such that

$$\max_{1\le i\le m} \sup_{s,t\in J_i} E|X_1(t) - X_1(s)| \le \varepsilon \tag{7.3.5}$$

where $J_i = [t_{i-1}, t_i)$. It can also be assumed that $t_i - t_{i-1} < \delta$ for all $1 \le i \le m$. For each n and $t \in [0,1]$

$$\begin{aligned} &|\tfrac{1}{n} \textstyle\sum_{k=1}^n X_k(t) - EX_1(t)| \\ &\le |\tfrac{1}{n} \textstyle\sum_{k=1}^n X_k(t) I_{[X_k \in K]} - EX_1(t) I_{[X_1 \in K]}| \\ &\quad + \tfrac{1}{n} \textstyle\sum_{k=1}^n \|X_k\| I_{[X_k \notin K]} + E(\|X_1\| I_{[X_1 \notin K]}). \end{aligned} \tag{7.3.6}$$

By Theorem 3.1.5

$$\frac{1}{n} \sum_{k=1}^n \|X_k\| I_{[X_k \notin K]} \to E(\|X_1\| I_{[X_1 \notin K]}) \tag{7.3.7}$$

with probability one. For the first term of (7.3.6),

$$\sup_{t\in J_i} |\tfrac{1}{n} \textstyle\sum_{k=1}^n X_k(t) I_{[X_k \in K]} - EX_1(t) I_{[X_1 \in K]}|$$

$$\leq \left|\frac{1}{n}\sum_{k=1}^{n} X_k(t_{i-1}) I_{[X_k \in K]} - EX_1(t_{i-1}) I_{[X_1 \in K]}\right|$$

$$+ \sup_{t \in J_i} \left|\frac{1}{n}\sum_{k=1}^{n} X_k(t) I_{[X_k \in K]} - \frac{1}{n}\sum_{k=1}^{n} X_k(t_{i-1}) I_{[X_k \in K]}\right|$$

$$+ \sup_{t \in J_i} \left|EX_1(t) I_{[X_1 \in K]} - EX_1(t_{i-1}) I_{[X_1 \in K]}\right|$$

$$\leq \left|\frac{1}{n}\sum_{k=1}^{n} X_k(t_{i-1}) I_{[X_k \in K]} - EX_1(t_{i-1}) I_{[X_1 \in K]}\right|$$

$$+ \frac{1}{n}\sum_{k=1}^{n} |X_k(t_i - 0) - X_k(t_{i-1})| I_{[X_k \in K]}$$

$$+ E(|X_1(t_i - 0) - X_1(t_{i-1})| I_{[X_1 \in K]}) + \varepsilon \tag{7.3.8}$$

by compactness of K and $t_i - t_{i-1} < \delta$. By the SLLN of Theorem 3.1.5

$$\left|\frac{1}{n}\sum_{k=1}^{n} X_k(t_{i-1}) I_{[X_k \in K]} - EX_1(t_{i-1}) I_{[X_1 \in K]}\right| \to 0 \tag{7.3.9}$$

with probability one for each $i = 1,\ldots,m$. Thus, from (7.3.7), (7.3.8), and (7.3.9)

$$\beta(\limsup_n \left\|\frac{1}{n}\sum_{k=1}^{n} X_k I_{[X_k \in K]} - EX_1 I_{[X_1 \in K]}\right\|_\infty)$$

$$\leq 2 \max_{1 \leq i \leq m} E(|X_1(t_i - 0) - X_1(t_{i-1})| I_{[X_1 \in K]}) + \varepsilon$$

$$\leq 2\varepsilon + \varepsilon = 3\varepsilon. \tag{7.3.10}$$

Thus from (7.3.6) and (7.3.10)

$$\beta(\limsup \left\|\frac{1}{n}\sum_{k=1}^{n} X_k - EX_1\right\|_\infty)$$

$$\leq 2E(\|X_1\| I_{[X_1 \notin K]}) + 3\varepsilon = 5\varepsilon.$$

The proof is completed by choosing a sequence of $\{\varepsilon_n\}$ converging to zero and by exlcuding the corresponding countable collection of null sets. ///

The concept of convex tightness will be used to obtain a strong law of large numbers for independent random elements which are not necessarily identically distributed.

Theorem 7.3.3 Let $\{X_n\}$ be a sequence of independent convex tight random elements in D satisfying $\sup_n E||X_n||_\infty^r \le \Gamma$, where $r > 1$ and Γ is a constant. Then

$$d(\tfrac{1}{n}\textstyle\sum_{k=1}^n X_k, \tfrac{1}{n}\sum_{k=1}^n EX_k) \to 0$$

with probability one.

Proof: Let $\varepsilon > 0$ be given. By convex tightness let $K \subset D$ be convex and compact such that $P[X_k \notin K] < \varepsilon^{\frac{r}{r-1}}$, for all k, and w.l.o.g., it can be assumed that $0 \in K$. Then

$$E||X_k I_{[X_k \notin K]}||_\infty \le (E||X_k||^r)^{1/r} \cdot P[X_k \notin K]^{\frac{r-1}{r}} \le \Gamma^{1/r}\varepsilon. \quad (7.3.11)$$

Note that EX_k exists for every k since $E||X_k||_\infty \le \Gamma + 1$. For each n

$$d(\tfrac{1}{n}\textstyle\sum_{k=1}^n X_k, \tfrac{1}{n}\sum_{k=1}^n EX_k)$$

$$= d(\tfrac{1}{n}\textstyle\sum_{k=1}^n (X_k I_{[X_k \in K]} + X_k I_{[X_k \notin K]}), \tfrac{1}{n}\sum_{k=1}^n (EX_k I_{[X_k \in K]} + EX_k I_{[X_k \notin K]}))$$

$$\le d(\tfrac{1}{n}\textstyle\sum_{k=1}^n X_k I_{[X_k \in K]}, \tfrac{1}{n}\sum_{k=1}^n T_m(X_k I_{[X_k \in K]})) \quad \text{(I)}$$

$$+ d(\tfrac{1}{n}\textstyle\sum_{k=1}^n T_m(X_k I_{[X_k \in K]}), \tfrac{1}{n}\sum_{k=1}^n T_m(EX_k I_{[X_k \in K]})) \quad \text{(II)}$$

$$+ d(\tfrac{1}{n}\textstyle\sum_{k=1}^n T_m(EX_k I_{[X_k \in K]}), \tfrac{1}{n}\sum_{k=1}^n EX_k I_{[X_k \in K]}) \quad \text{(III)}$$

$$+ ||\tfrac{1}{n}\textstyle\sum_{k=1}^n X_k I_{[X_k \notin K]}||_\infty \quad \text{(IV)}$$

$$+ ||\tfrac{1}{n}\textstyle\sum_{k=1}^n EX_k I_{[X_k \notin K]}||_\infty. \quad \text{(V)}$$

The above inequality is a consequence of Lemma 7.1.4.

For (I),

$$d(\tfrac{1}{n}\textstyle\sum_{k=1}^n X_k I_{[X_k \in K]}, T_m(\tfrac{1}{n}\sum_{k=1}^n X_k I_{[X_k \in K]})) \le \sup_{x \in K} d(x, T_m(x)), \quad (7.3.12)$$

pointwise in Ω, since K is convex and $0 \in K$. Since $\limsup_m \sup_{x \in K} d(x, T_m(x)) = 0$ from Corollary 7.1.8, there is m_o such that (I) $< \varepsilon$ for all $m \geq m_o$, and every sample point $\omega \in \Omega$.

Using the linearity of T_m,

$$(II) \leq ||T_m(\tfrac{1}{n} \textstyle\sum_{k=1}^n (X_k I_{[X_k \in K]} - E[X_k I_{[X_k \in K]}]))||_\infty$$

$$\textstyle\sum_{i=1}^{2^m} |\tfrac{1}{n} \sum_{k=1}^n (X_k(\tfrac{i}{2^m}) I_{[X_k \in K]} - E[X_k(\tfrac{i}{2^m}) I_{[X_k \in K]}])|. \qquad (7.3.13)$$

Since K is compact, there is a constant C such that $\sup_{x \in K} \sup_{0 \leq t \leq 1} |x(t)| \leq C$. For each $i = 1, \ldots, 2^m$, $|X_k(\frac{i}{2^m}) I_{[X_k \in K]}| \leq C$, and hence, by the strong law of large numbers for random variables [Theorem 3.1.4]

$$\lim_{n \to \infty} \frac{1}{n} \textstyle\sum_{k=1}^n (X_k(\tfrac{i}{2^m}) I_{[X_k \in K]} - E[X_k(\tfrac{i}{2^m}) I_{[X_k \in K]}]) = 0$$

with probability one. Thus, for each given m, almost surely (II) $< \varepsilon$ for all sufficiently large n.

By Corollary 7.1.8, $\lim_{m \to \infty} \sup_n$ (III) $= 0$. (7.3.14)

Next,

$$(IV) \leq \frac{1}{n} \textstyle\sum_{k=1}^n ||X_k||_\infty I_{[X_k \notin K]}$$

$$= \frac{1}{n} \textstyle\sum_{k=1}^n [\,||X_k||_\infty I_{[X_k \notin K]} - E||X_k||_\infty I_{[X_k \notin K]}]$$

$$+ \frac{1}{n} \textstyle\sum_{k=1}^n E||X_k||_\infty I_{[X_k \notin K]}. \qquad (7.3.15)$$

For the first of these terms it follows that

$$\textstyle\sum_{k=1}^\infty \frac{1}{k^r} E|\,||X_k||_\infty I_{[X_k \notin K]} - E||X_k||_\infty I_{[X_k \notin K]}|^r$$

$$\leq \textstyle\sum_{k=1}^\infty \frac{2^{r-1}}{k^r} \{E||X_k I_{[X_k \notin K]}||^r + (E||X_k||_\infty I_{[X_k \notin K]})^r\}$$

$$\le 2^r \sum_{k=1}^{\infty} \frac{1}{k^r} E||X_k||_\infty^r I_{[X_k \notin K]} \le 2^r \sum_{k=1}^{\infty} \frac{\Gamma}{k^r} < \infty,$$

since $r > 1$. Hence by K. L. Chung's strong law of large numbers, the first term tends to zero with probability one as $n \to \infty$. For the second term, (7.3.11) yields

$$\frac{1}{n} \sum_{k=1}^{n} E||X_k||_\infty I_{[X_k \notin K]} < \varepsilon \Gamma^{1/r} \tag{7.3.16}$$

for every n.

Finally, by (7.3.11), (V) $\le \Gamma^{1/r}\varepsilon$.

Thus, a null set can be excluded for each m, and the countable union Ω_0 is obtained. For $\varepsilon > 0$ and $\omega \notin \Omega_0$, m is chosen large enough so that (I) and (III) are each less than ε. Then $N(\varepsilon,\omega)$ is chosen large enough so that (II) and (IV) are each less than ε.///

The following strong law of large numbers by Daffer and Taylor (1977) can be used for the random elements which were excluded from Theorem 7.3.3 by Example 7.1.5. Let $D\uparrow$ denote the cone of non-decreasing elements of D.

Theorem 7.3.4 Let $\{X_n\}$ be a sequence of independent random elements in D satisfying

(i) $X_n \in D\uparrow$ almost surely, for each n;

(ii) $\sum \frac{E||X_n||_\infty^r}{n^r} < \infty$ for some $1 \le r \le 2$;

(iii) $EX_n = EX_1$, for all n.

Then,

$$||\frac{1}{n} \sum_{k=1}^{n} X_k - EX_1||_\infty \to 0$$

with probability one.

Proof: Put $\overline{X}_n = \frac{1}{n}\sum_1^n X_k$ and $\mu = EX_1$. Note that $E||X_1||_\infty^r < \infty$ implies the existence of μ. By Billingsley (1968), page 110, Lemma 1, for each m there is a partition of $[0,1]$, $0 = t_0 < t_1 < \dots < t_{k(m)} = 1$ such that $\sup_{t_i \le t, s \le t_{i+1}} |\mu(s) - \mu(t)| < 1/m$, for $i = 0,1,\dots,k(m) - 1$. Since $\mu \in D\uparrow$ this means $\mu(t_{i+1} - 0) - \mu(t_i) < 1/m$, for $i = 0,1,\dots,k(m) -1$. Let $t \in [0,1]$, then $t \in [t_{i-1}, t_i)$ for some $i = 1,2,\dots,k(m)$ or $t = 1$. In any case,

$$\begin{aligned}\overline{X}_n(t) - \mu(t) &\le \overline{X}_n(t_i) - 0) - \mu(t_{i-1}) \\ &\le \overline{X}_n(t_i - 0) - \mu(t_i - 0) + 1/m \qquad (7.3.17)\end{aligned}$$

and

$$\begin{aligned}\overline{X}_n(t) - \mu(t) &\ge \overline{X}_n(t_{i-1}) - \mu(t_i - 0) \\ &\ge \overline{X}_n(t_{i-1}) - \mu(t_{i-1}) - 1/m \qquad (7.3.18)\end{aligned}$$

Thus,

$$|\overline{X}_n(t) - \mu(t)| \le \max\{|\overline{X}_n(t_i-0)-\mu(t_i-0)|, |\overline{X}_n(t_{i-1})-\mu(t_{i-1})|\}+1/m$$

$$\le \max_{1\le i\le k(m)} [\max\{|\overline{X}_n(t_i-0)-\mu(t_i-0)|, |\overline{X}_n(t_{i-1})-\mu(t_{i-1})|\}]+1/m$$

$$\to 0 + 1/m$$

with probability one by the SLLN using Chung's condition of (ii). Thus,

$$\lim_{n\to\infty} ||\overline{X}_n - \mu||_\infty \le 1/m$$

with probability one, and since m is arbitrary,

$$||\overline{X}_n - \mu||_\infty \to 0$$

with probability one. ///

Remark: The conclusion immediately implies convergence in the Skorohod metric. A corresponding theorem holds mutatis mutandis for almost surely non-increasing random elements.

Weak laws of large numbers can be obtained in a similar manner. However, the hypotheses for the weak laws of large numbers involve less restrictive conditions, for example, pointwise conditions will often suffice.

7.4 A DISCUSSION OF CONVEX TIGHTNESS

In this section some characterizations of convex tightness are presented. The general appearance of these characterizations are mainly "finite-dimensional" or "small" in $|| \; ||_\infty$.

Roughly speaking, convex compact sets in D are those sets of functions which become arbitrarily small in absolute value at cluster points of jumps in [0,1]. The following theorem characterizes the convex compact subsets K of D in terms of the jumps of functions in K.

For $A \subset D$ and $\varepsilon > 0$, let

$$S_\varepsilon(A) = \{t \in [0,1]: \sup_{x \in A}|x(t) - x(t-0)| > \varepsilon\}. \quad (7.4.1)$$

Also, let co(A) denote the convex hull of A.

Theorem 7.4.1 If K is a relatively compact subset of D, then co(K) is relatively compact if and only if $S_\varepsilon(K)$ is finite for every $\varepsilon > 0$.

Proof ("only if" part): Suppose that for some $\varepsilon > 0$, $S_\varepsilon(K)$ is infinite. Then there is $t_o \in [0,1]$ and a sequence $\{t_n\}$ of distinct points in $S_\varepsilon(K)$ converging to t_o and a sequence $\{x_n\}$ of elements of K such that $|x_n(t_n) - x_n(t_n-0)| > \varepsilon$ for each n. Since

K is compact, there is a $\delta > 0$ such that

$$\sup_{x \in K} w_x'(\delta) < \varepsilon/2 \tag{7.4.2}$$

Find n and n', n ≠ n', such that $t_n, t_{n'} \in (t_0-\delta/2,\ t_0+\delta/2)$ and such that $|x_n(t_n) - x_n(t_n-0)| > \varepsilon$ and $|x_{n'}(t_{n'}) - x_{n'}(t_{n'}-0)| > \varepsilon$. Since $x_n, x_{n'}$ have jumps at $t_n, t_{n'}$, respectively, of magnitudes $> \varepsilon$, it follows from (7.4.2) that for $i = n, n'$

$$\sup_{t_i \le t, s \le t_i+\delta} |x_n(t) - x_n(s)| < \varepsilon/2$$

and

$$\sup_{t_i-\delta \le t, s < t_i} |x_n(t) - x_n(s)| < \varepsilon/2. \tag{7.4.3}$$

Now let $T = \{t_0, t_1, \ldots, t_m\}$ be any partition of [0,1] with $\min_{1\le i\le m} \{t_i - t_{i-1}\} \ge \delta$. Since $|t_n - t_{n'}| < \delta$, T can contain t_n or $t_{n'}$ or neither one, but not both. If for example $t_n \in T$, then $t_{n'} \notin T$ and

$$\begin{aligned} &|x_n(t_{n'}) + x_{n'}(t_{n'}) - (x_n(t_{n'}-0) + x_{n'}(t_{n'}-0))| \\ &\ge |x_{n'}(t_{n'}) - x_{n'}(t_{n'}-0)| - |x_n(t_{n'}) - x_n(t_{n'}-0)| \\ &> \varepsilon - \varepsilon/2 = \varepsilon/2 \end{aligned}$$

using the relations (7.4.3).

Define $x_\delta = \frac{1}{2}x_n + \frac{1}{2}x_{n'}$. Then, $|x_\delta(t_{n'}) - x_\delta(t_{n'}-0)| > \varepsilon/4$. If $t_{n'} \in T$, then $t_n \notin T$ and the same reasoning as above yields $|x_\delta(t_n) - x_\delta(t_n-0)| > \varepsilon/4$. If $t_n \notin T$ and $t_{n'} \notin T$, then a fortiori $|x_\delta(t_n) - x_\delta(t_n-0)| > \varepsilon/4$. Thus, for any partition T with $\min_{1\le i\le m} \{t_i - t_{i-1}\} \ge \delta$,

$$\sup_{s,t \in [t_{i-1}, t_i)} |x_\delta(s) - x_\delta(t)| > \varepsilon/4$$

for some $i = 1,\ldots,m$, and hence $w'_{x_\delta}(\delta) \geq \varepsilon/4$. But $x_\delta \in co(K)$, and so $\sup_{x \in co(K)} w'_x(\delta) \geq w'_{x_\delta}(\delta) \geq \varepsilon/4$.

Since $\varepsilon > 0$ is fixed and $\delta > 0$ is arbitrary, this yields $\underline{\lim}_{\delta \to 0} \sup_{x \in co(K)} w'_x(\delta) > 0$, and thus $co(K)$ is not relatively compact in D. ///

For the "if" part we first prove the following lemma.

Lemma 7.4.2 If K is a relatively compact subset of D such that $S_\varepsilon(K)$ is finite for each $\varepsilon > 0$, then to each $t \in [0,1]$ the following holds: for each $\varepsilon > 0$ there exists $\delta > 0$ such that $\sup_{x \in K} w_x([t,t+\delta)) < \varepsilon$ and $\sup_{x \in K} w_x((t-\delta,t)) < \varepsilon$.

Proof: Let $\varepsilon > 0$ be given and fix $t_o \in [0,1]$. By relative compactness of K, find $\delta_o > 0$ such that $\sup_{x \in K} w'_x(\delta_o) < \varepsilon/3$. Now find δ_1, $0 < \delta_1 \leq \delta_0$, such that $t \in (t_0, t_0 + \delta_1)$ implies

$$\sup_{x \in K} |x(t) - x(t-0)| < \varepsilon/3. \tag{7.4.4}$$

Take $x \in K$ and let $T = \{t_i\}$ be any finite partition of $[0,1]$ such that $\max_i \sup_{x \in K} w_x([t_{i-1},t_i)) < \varepsilon/3$ and $\min_i \{t_i - t_{i-1}\} \geq \delta_0$. If no point of T falls in $[t_0, t_0 + \delta_1)$, then $w_x([t_0,t_0 + \delta_1)) < \varepsilon/3$. If a point $t_i \in T$ is such that $t_i \in (t_0, t_0 + \delta_1)$, then $\max_i \sup_{x \in K} w_x([t_{i-1},t_i)) < \varepsilon/3$ yields $w_x([t_0,t_i)) < \varepsilon/3$ and $w_x([t_i, t_0 + \delta_1)) < \varepsilon/3$. By Inequality (7.4.4), if x makes a jump at t_i, its magnitude is necessarily $< \varepsilon/3$. Thus, by the triangle inequality, $w_x([t_0,t_0 + \delta_1)) < \varepsilon/3 + \varepsilon/3 + \varepsilon/3 = \varepsilon$, for any $x \in K$. Hence, $\sup_{x \in K} w_x([t_0,t_0 + \delta_1)) \leq \varepsilon$. In a similar manner, a $\delta_2 > 0$ can be found such that $\sup_{x \in K} w_x((t_0-\delta_2, t_0)) \leq \varepsilon$. Now take $\delta = \min\{\delta_1,\delta_2\}$ and the proof is complete. ///

Proof ("if" part): Now, let $\varepsilon > 0$ be given and $S_\varepsilon(K)$ be finite. Let $S_\varepsilon(K) = \{t_1, \ldots, t_N\}$.

Let $I_k = [t_k, t_{k+1}]$ for $k = 0,1,\ldots,N$, where $t_o = 0$ and $t_{N+1} = 1$. Put $t_{k1} = t_k$ and corresponding to t_k find, by Lemma 7.4.2, $\delta(t_{k1}) > 0$ such that

$$\sup_{x \in K} w_x([t_k, t_k + \delta(t_{k1}))) < \varepsilon. \tag{7.4.5}$$

For $j \geq 1$, inductively define $t_{k,j+1} = t_{kj} + \delta(t_{kj})$ and set $I_{k1} = [t_{k1}, t_{k1} + \delta(t_{k1}))$ and $I_{kj} = (t_{kj}, t_{kj} + \delta(t_{kj}))$ if $j \neq 1$. Let t_{kj} and $\delta(t_{kj})$ be determined alternately in the following manner: given $t_{k1}, \ldots, t_{kj}$, find by Lemma 7.4.2, $\delta(t_{kj}) > 0$, such that

$$\sup_{x \in K} w_x((t_{k,j+1} - \delta(t_{k,j+1}), t_{k,j+1})) < \varepsilon$$

and

$$\sup_{x \in K} w_x([t_{k,j+1}, t_{k,j+1} + \delta(t_{k,j+1}))) < \varepsilon.$$

Each $t_{kj} \leq t_{k+1}$ since some $x \in K$ makes a jump at t_{k+1}.

In this way a sequence $\{t_{k1}, t_{k2}, \ldots\}$ of points in $[t_k, t_{k+1}]$ is obtained, and a sequence of intervals $I_{k1}, I_{k2}, \ldots$ which are all open sets in $[t_k, t_{k+1}]$ is obtained. Another application of Lemma 7.4.2 yields a $\delta(t_{k+1}) > 0$ such that

$$\sup_{x \in K} w_x((t_{k+1} - \delta(t_{k+1}), t_{k+1})) < \varepsilon.$$

Let $I_k' = (t_{k+1} - \delta(t_{k+1}), t_{k+1}]$.

The collection of relatively open subintervals $\{I_k', I_{k1}, I_{k2}, \ldots\}$ is an open cover of $[t_k, t_{k+1}]$ which is compact. Consequently, there exists an open subcover $\{J_{k1}, \ldots, J_{kN_k}\}$. Let $\{s_{k1}, \ldots, s_{kN_k}\}$ denote the respective centers of these intervals.

Now this can be done for every k, k = 0,1,...,N. The collection of points $\bigcup_{k=0}^{N} \bigcup_{j=1}^{N_k} \{s_{kj}\} \cup \bigcup_{j=1}^{N} \{t_j\}$ forms a partition of [0,1], call it T, and denote the points of it in ascending order, by

$$0 = s_0 < s_1 < \ldots < s_{m-1} < s_m = 1.$$

The claim is now that

$$\max_{i=1,\ldots,m-1} \sup_{x \in co(K)} w_x([s_{i-1}, s_i)) < \varepsilon \tag{7.4.6}$$

and

$$\sup_{x \in co(K)} w_x([s_{m-1}, 1]) < \varepsilon.$$

Let $x = \sum_{j=1}^{n} \alpha_j x_j$, $x_j \in K$, $\alpha_j \geq 0$ and $\sum_{j=1}^{n} \alpha_j = 1$.

Then,
$$\begin{aligned}
& w_x([s_{i-1}, s_i)) \\
&= \sup_{s,t \in [s_{i-1}, s_i)} \left| \sum_{j=1}^{n} \alpha_j (x_j(s) - x_j(t)) \right| \\
&\leq \sum_{j=1}^{n} \alpha_j \sup_{s,t \in [s_{i-1}, s_i)} |x_j(s) - x_j(t)| \\
&= \sum_{j=1}^{n} \alpha_j w_{x_j}([s_{i-1}, s_i)) \\
&\leq \sum_{j=1}^{n} \alpha_j \sup_{x \in K} w_x([s_{i-1}, s_i)) \\
&\leq \sum_{j=1}^{n} \alpha_j \varepsilon = \varepsilon.
\end{aligned}$$

Hence, (7.4.6) holds and thus

$$\sup_{x \in co(K)} w'_x(\delta) < \varepsilon$$

by taking $\delta < \min_{1 \leq i \leq m} \{s_i - s_{i-1}\}$.

Thus, to each $\varepsilon > 0$ there is $\delta > 0$ such that $\sup_{x \in co(K)} w_x'(\delta) < \varepsilon$, and so

$$\lim_{\delta \to 0} \sup_{x \in \overline{co(K)}} w_x'(\delta) = 0$$

and co(K) is relatively compact. ///

Finally, the weighted sums results of Chapter V could be proved for D[0,1] using the techniques of this chapter. However, it is hoped that results less restrictive than those for convex tightness can be developed first. In addition, a closer examination of possible applications in D[0,1] should be considered.

7.5 PROBLEMS

7.1 Let K be convex and relatively compact. Show that the convex hull of $\{0\} \cup K$ is relatively compact.

7.2 When $EX_1 \in C[0,1] \subset D[0,1]$ in Theorem 7.2.2, show that

$$\left|\left|\frac{1}{n} \sum_{k=1}^{n} X_k - EX_1\right|\right|_\infty \to 0$$

with probability one.

7.3 In Lemma 7.3.1 show that $E|X(1) - X(0)| \leq \varepsilon$ when $\tau_1 = 1$.

7.4 Let the random element $X \in D\uparrow$ with probability one. Show that $EX \in D\uparrow$ when $E||X||_\infty < \infty$.

7.5 Prove that scalar multiplication is a continuous operation in D[0,1] with the Skorohod topology. Also, multiplication is closed in D[0,1] but is not a continuous operation.

CHAPTER VIII

POSSIBLE APPLICATIONS

8.0 INTRODUCTION

Applications of the laws of large numbers for random variables to various problems in statistics, information theory, and number theory is well-documented. This chapter will list some possible applications of the stochastic convergence results of Chapters IV, V, and VI. While particular techniques for specific applications often provide stronger results, the adaptation of the general function space techniques will provide powerful tools for many problems. Thus, several possible applications will be indicated in this chapter.

Example 4.2.1 illustrated how easily Brownian motion processes fit the random elements framework. Example 6.2.1 indicated the importance of allowing the weights to random in regression technique. Of course, the applications of the results of Chapter VII for D[0,1] is documented by Billingsley (1968) and many others.

Since the area of stochastic processes provided the motivation for the study of random elements in linear topological spaces, many of the applications given will be concerned with stochastic processes. Section 8.1 will include applications for Brownian motion processes and uniform convergence of some discrete parameter processes. An interesting application in statistical decision theory will be indicated in Section 8.2. Section 8.3 will discuss the use of weighted sum results in control charts, and Section 8.4 will contain some estimation problems. In particular, an approach to density estimation will be discussed.

8.1 APPLICATIONS IN STOCHASTIC PROCESSES

In this section some of the results of Chapters IV, V, and VI will be applied to stochastic processes which can be regarded as random elements in the particular spaces $C[0,1]$, c_0, ℓ^p, and $C[0,\infty)$.

To indicate a possible application for the tightness results, consider a manufacturing problem where items are produced by different processes (assembly lines, machines, operators, etc.). At the start of each day the settings are exact. During the working day, various factors (temperature, wear, etc.) change the output in a continuous manner which is approximately Brownian motion with a drift, $d(t)$. The drift for each process (of interest) can be assumed to be the same but not the variances. If $d(t)$ could be estimated, then an overall continuous adjusting during the day could minimize the percent out of specifications.

Samples of size n could be taken during the day and the average could be plotted.

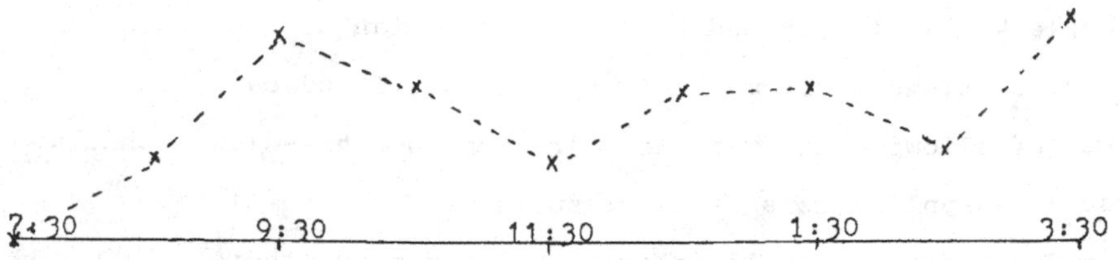

By connecting the sample means polygonally for the different times, an estimate

$$g_8(t) = \frac{1}{n} \sum_{i=1}^{n} U_8(t) = \frac{1}{n} \sum_{i=1}^{n} U_8(V_i(t)) \qquad (8.1.1)$$

of the drift $d(t)$ can be obtained. In (8.1.1), $\{V_i(t): t \in [0,8]\}$ denotes the Brownian motion processes and U_8 is the partial sum operator for the Schauder basis which is given in Example 1.3.2.

Let V_0 denote a Brownian motion process with $E[V_0(1)^2] = 1$. For each $\varepsilon > 0$ there exists a compact set K_ε of $C[0,1]$ such that

$P[V_0 \in K_\varepsilon] > 1 - \varepsilon$, and without loss of generality K_ε can be assumed to be convex and symmetric. Let $\sigma_{V_i}^2 = E[V_i(1)^2] - (d(1))^2$, and assume that the sequence $\{E[V_i(1)^2]\}$ is a bounded sequence (bounded by M). Then,

$$1 - \varepsilon < P[V_0 \in K_\varepsilon] = P[V_i - d(t) \in (\sigma_{V_i})K_\varepsilon]$$
$$= P[V_i - d(t) \in MK_\varepsilon]$$

for each i. Thus, the independent random elements $\{V_i - d(t)\}$ in C[0,8] are tight and their second absolute moments are uniformly bounded. Corollary 5.2.8 implies that

$$||\tfrac{1}{n} \sum_{i=1}^n V_i - EV_1|| = \sup_{0 \le t \le 8} |\tfrac{1}{n} \sum_{i=1}^n V_i(t) - d(t)| \to 0$$

with probability one. Also, for each m

$$\sup_{0 \le t \le 8} |\tfrac{1}{n} \sum_{i=1}^n U_m(V_i(t)) - U_m(d(t))| \to 0$$

with probability one since $U_m(d(t)) = U_m(EV_1) = E(U_mV_1)$. Interestingly $E(U_mV_1) = E(\frac{1}{n} \sum_{i=1}^n U_m(V_i)) = E(g_m(t))$ as constructed in (8.1.1). Since, $U_m(d(t)) \to d(t)$ as $m \to \infty$, a uniform estimate of the drift is assured, and the rates of convergence can be calculated.

Similar examples of uniform convergence in probability for Brownian motion processes on $[0,\infty)$ can be constructed by using the weak laws of large numbers for $C[0,\infty)$. In these applications,

$$\text{Cov}\ (V_n(t) - V_n(s), V_k(t) - V_k(s)) = 0$$

for all $t > s$ and $k \neq n$ is sufficient for the random elements to be weakly uncorrelated. While these applications for $C[0,\infty)$ will not be discussed, it is important to note that the less restrictive coordinate uncorrelation suffices in applying the weak laws of large numbers.

The space of null convergent sequences, c_0, is used for the next application. Let $\{V_n\}$ be a sequence of stochastic processes where each process has parameter space $\{1,2,\ldots\}$. Also, for each n and k let the stochastic processes V_n and V_k have the same finite-dimensional distributions and let $\lim_{m\to\infty} V_n(m) = 0$ with probability one for each n. Finally, let $E\|V_1\| = E[\sup_m |V_1(m)|] < \infty$ and for each m let the weak law of large numbers hold for the random variables $\{V_n(m): n \geq 1\}$. From the results of Chapter II $\{V_n\}$ can be regarded as a sequence of random elements in c_0 with $EV_1 = (EV_1(1), EV_1(2),\ldots)$. Moreover, the random elements are identically distributed since they have the same finite-dimensional distributions [from Property 2.3.4]. Theorem 4.1.3 states that for any $\varepsilon > 0$

$$P[\|\tfrac{1}{n}\textstyle\sum_{k=1}^n V_k - EV_1\| > \varepsilon]$$

$$= P[\sup_m |\tfrac{1}{n}\textstyle\sum_{k=1}^n V_k(m) - EV_1(m)| > \varepsilon] \to 0. \qquad (8.1.2)$$

One consequence of (8.1.2) is the uniform convergence of $\frac{1}{n}\sum_{k=1}^n V_k(m)$ to $EV_1(m)$ in probability.

In essence, the preceding paragraph gives a uniform weak law of large numbers for triangular arrays of random variables. Let $\{V_{nm}: n \geq 1, m \geq 1\}$ be a family of random variables satisfying the following conditions:

(i) $\lim_{m\to\infty} V_{nm} = 0$ with probability one for each n; (8.1.3)

(ii) for each m the weak law of large numbers holds for the random variables $\{V_{nm}: n \geq 1\}$; (8.1.4)

(iii) for any n and any $j \geq 1$ the stochastic processes $\{V_{nm}: m \geq 1\}$ and $\{V_{(n+j)m}: m \geq 1\}$ have the same finite-dimensional distributions; and (8.1.5)

(iv) $E[\sup_m |V_{1m}|] < \infty$. (8.1.6)

It follows from the above discussion that $\frac{1}{n}\sum_{k=1}^n V_{km} \to EV_{1m}$ in probability uniformly for m. Thus,

$$\left|\frac{1}{n}\sum_{k=1}^{n} V_{kn} - EV_{1n}\right| \le \sup_m \left|\frac{1}{n}\sum_{k=1}^{n} V_{km} - EV_{1m}\right|$$
$$= \left\|\frac{1}{n}\sum_{k=1}^{n} V_k - EV_1\right\| \to 0$$

in probability.

Consider now the space ℓ^1. Let $\{V_n(m)\colon\ m \ge 1,\ n \ge 1\}$ be a family of random variables satisfying the following conditions:

(i) $\sum_{m=1}^{\infty} E|V_1(m)| < \infty$ and $\sum_{m=1}^{\infty} |V_n(m)| < \infty$ with probability one for each n; (8.1.7)

(ii) the stochastic processes $\{V_n(m)\colon\ m \ge 1\}$ and $\{V_{n+j}(m)\colon\ m \ge 1\}$ have the same finite-dimensional distributions for each $n \ge 1$ and $j \ge 1$; and (8.1.8)

(iii) for each $m \ge 1$ $\{V_n(m)\colon\ n \ge 1\}$ satisfies the weak law of large numbers. (8.1.9)

Again, $\{V_n = (V_n(1), V_n(2), \ldots)\colon\ n \ge 1\}$ can be regarded as a sequence of identically distributed random elements in the space ℓ^1 by (8.1.7) and (8.1.8) and Chapter II. Condition (8.1.7) implies that $E\|V_1\| < \infty$ and hence that $EV_1 = (EV_1(1), EV_1(2), \ldots)$ exists. From Theorem 4.1.3 it follows that for any $\varepsilon > 0$

$$P\left[\left\|\frac{1}{n}\sum_{k=1}^{n} V_k - EV_1\right\| > \varepsilon\right] \to 0. \tag{8.1.10}$$

But, (8.1.10) can be written as

$$P\left[\sum_{m=1}^{\infty} \left|\frac{1}{n}\sum_{k=1}^{n} V_k(m) - EV_1(m)\right| > \varepsilon\right] \to 0. \tag{8.1.11}$$

Since the spaces $\ell^p (p > 1)$ can be handled similarly, this application also gives a weak law of large numbers for random elements in these spaces.

For notational convenience the laws of large numbers results were used. The techniques in applying the convergence of weighted sums are similar. Applications in Fréchet spaces also follow easily using the same techniques.

8.2 APPLICATIONS IN DECISION THEORY

In standard statistical decision making procedures the selection of alternatives is a function of the sample data. An important property of statistical decision making procedures is convergence of the alternative function to the "best" alternative as the sample size tends to infinity. In this section the standard statistical decision making procedures will be extended to multiattributed decision problems, and the convergence of the alternative function to the "best" alternative will be proved using the weak laws of large numbers which were developed in Chapter IV. Two particular examples of statistical decision problems in infinite-dimensional spaces will be given, one of which has direct applications in sequential sampling.

Let X denote the set of all possible alternatives which are available to man and let Θ denote the set of all possible states of nature. Utility is defined as a measure of satisfication and is given by $U(\theta,x)$, where $U(\theta,x)$ represents the amount of satisfaction for man using alternative x when the state of nature is θ. In applications the determination of the utility function $U(\theta,x)$ is a major problem. In this discussion the construction in Lindgren (1963, pages 153-4) will be used to assume that utility is measured by a loss function $L(\theta,x)$. The loss function represents the "loss" suffered by man when alternative x is used and the actual state of nature is θ.

If a random sample can be obtained where the parameters of the probability distribution involve the actual state of nature θ, then statistical decision procedures can be used in choosing alternative x.

Thus, the selection of alternatives becomes a function $x(Z_n)$ of the random sample $Z_n = (V_1,\ldots,V_n)$.

A statistical procedure $x(Z_n)$ for choosing an alternative x using a sample Z_n is said to be consistent if for each θ the expected value of the loss function, $E[L(\theta,x(Z_n))]$, converges to the greatest lower bound for $L(\theta,x)$ as the sample size n goes to ∞. A statistical procedure $x(Z_n)$ is said to be consistent with convergence in probability if for each θ, $L(\theta,x(Z_n))$ converges in probability to the greatest lower bound for $L(\theta,x)$ as n goes to ∞.

In addition to the results of Chapter IV, the following theorem can be used to develop consistent statistical decision procedures. The proof is similar to the proofs in Section 4.1 and is omitted.

Theorem 8.2.1 Let Y be a Banach space which has a Schauder basis or a normed linear space which has a monotone basis. Let $\{V_n: n = 1,2,\ldots\}$ be a sequence of random elements in Y which are coordinatewise identically distributed and such that there exists a k_0 with $E||Q_m(V_n)|| \le E||Q_m(V_{k_0})||$ and $E||V_{k_0}|| < \infty$, where Q_m is the linear operator defined in Section 1.3. For each coordinate functional f_m

$$\left|\frac{1}{n}\sum_{k=1}^n f_m(V_k) - Ef_m(V_1)\right| \to 0$$

in probability if and only if

$$\left|\left|\frac{1}{n}\sum_{k=1}^n V_k - EV_1\right|\right| \to 0$$

in probability.

Consider the problem of estimating θ_m by choosing x_m based on a random sample $\{V_1^{(m)},\ldots,V_n^{(m)}\}$ for each coordinate. Since $\{V_1^{(m)},\ldots,V_n^{(m)}\}$ is a random sample for each m, the random variables $V_1^{(m)},\ldots,V_n^{(m)}$ are independent and identically distributed. Hence,

$\{V_n = (V_n^{(1)},\ldots,V_n^{(m)},\ldots)\colon\ n = 1,2,\ldots\}$ is a sequence of random elements (in some sequence space which has a norm) that are at least coordinatewise independent. Moreover, if $EV_1^{(m)} = \theta_m$ for each m, then

$$EV_1 = (EV_1^{(1)},\ldots,EV_1^{(m)},\ldots) = (\theta_1,\ldots,\theta_m,\ldots) = \theta$$

by use of an appropriate basis. Let the alternative function based on the random sample $Z_n = (V_1,\ldots,V_n)$ be defined coordinatewise by

$$x(Z_n) = (\tfrac{1}{n}\textstyle\sum_{k=1}^{n} V_k^{(1)},\ldots,\tfrac{1}{n}\sum_{k=1}^{n} V_k^{(m)},\ldots) = \tfrac{1}{n}\sum_{k=1}^{n} V_k.$$

If the loss function $L(\theta,x)$ is a monotone increasing, continuous function of $||\theta - x||$, then Theorem 4.1.3 or Theorem 8.2.1 implies that the alternative function is consistent with convergence in probability if $E||V_{k_0}|| < \infty$ and $E||Q_m(V_n)|| \le E||Q_m(V_{k_0})||$ since

$$\tfrac{1}{n}\textstyle\sum_{k=1}^{n} V_k^{(m)} \to \theta_m$$

in probability for each m. Moreover, if $L(\theta,x)$ is assumed to be bounded for each $\theta \in \Theta$, then

$$X(Z_n) = \tfrac{1}{n}\textstyle\sum_{k=1}^{n} V_k$$

is also consistent. The moment conditions depend on the particular normed linear space which is used and the random sample Z_n but is often easily satisfied.

The following examples further illustrate these results.

Let each $\theta \in \Theta$ be of the form $\theta = (\theta_1,\ldots,\theta_m,0,\ldots)$, where the number of nonzero real numbers θ_i may vary with each θ. Also let each alternative $x \in X$ be of the form $x = (x_1,\ldots,x_m,0,\ldots)$, where each real number x_i is an estimate of θ_i. The norm can be a function of the error in each estimation. But since the errors in estimating each θ_i may not be of equal importance, let

$$||\theta - x|| = \sup_k w_k|\theta_k - x_k|, \tag{8.2.1}$$

where the w_k's are positive weights which can be assigned by the decision-maker. The different costs of adjustment in production is an example where the errors may be of unequal importance.

By the above construction Θ and X are subsets of the normed linear space $R^{(\infty)} = \{a = (a_1,\ldots,a_n,\ldots)\colon\ a_n = 0$ for all but a finite number of n's$\}$ where the norm of each $a \in R^{(\infty)}$ is $||a|| = \sup_k w_k|a_k|$. The observations are given by the random sample $Z_n = (V_1,\ldots,V_n)$, where each $V_k = (V_k^{(1)},\ldots,V_k^{(m)},\ldots)$ is a sequence of random variables such that $V_k \in R^{(\infty)}$ for each outcome. If $EV_k^{(m)} = \theta_m$ for each m, then an obvious choice for the alternative function is

$$x(Z_n) = \frac{1}{n}\sum_{k=1}^n V_k = \left(\frac{1}{n}\sum_{k=1}^n V_k^{(1)},\ldots,\frac{1}{n}\sum_{k=1}^n V_k^{(m)},\ldots\right). \tag{8.2.2}$$

Moreover, this alternative function is consistent with convergence in probability since Theorem 8.2.1 implies that $||\frac{1}{n}\sum_{k=1}^n V_k - \theta||$ converges to 0 in probability for each θ since the loss function $L(\theta,x)$ is a continuous, monotone increasing function of $||\theta - x||$. In addition, if $L(\theta,x)$ is bounded above for each θ, then $x(Z_n)$ is also consistent.

As a particular example consider sequential decision procedures. Recall that in these procedures decisions can be made that additional samples are to be taken. All the data can be used at each stage, and additional random samples $Z_n^{(m)} = \{V_1^{(m)},\ldots,V_n^{(m)}\}$ can be taken consecutively until convincing data are accumulated. When the sequential procedure terminates with probability one, then $\{V_k = (V_k^{(1)},\ldots,V_k^{(m)},\ldots)\colon\ 1 \le k \le n\}$ are random elements in $R^{(\infty)}$. The random elements are coordinatewise independent since each coordinate represents one of the random samples but are obviously not independent since the random amples can depend on each other. Thus, Theorem 8.2.1 provides for the uniform convergence in each coordinate (regardless of the weightings), and hence consistency for sequential decision procedures.

For another example let each state of nature $\theta = (\theta_1,\ldots,\theta_m,\ldots)$ and each alternative $x = (x_1,\ldots,x_m,\ldots)$ be sequences of real numbers. Again, the data can be coded so as to emphasize the relative importance of predicting θ_m by x_m for each m. For example, it may be assumed that Θ and X are subsets of the Banach space $\ell^1 = \{a = (a_1,\ldots,a_m,\ldots): \sum|a_m| < \infty\}$ and that a norm can be defined by $||\theta - x|| = \sum|\theta_m - x_m|$. The random sample is a set of infinite-dimensional random vectors $V_1,\ldots,V_n$, where each $V_k = (V_k^{(1)},\ldots,V_k^{(m)},\ldots) \in \ell^1$. If $E(V_k^{(m)}) = \theta_m$ for each m, then the alternative function can be defined as

$$x(Z_n) = \frac{1}{n}\sum_{k=1}^n V_k = \left(\frac{1}{n}\sum_{k=1}^n V_k^{(1)},\ldots,\frac{1}{n}\sum_{k=1}^n V_k^{(m)},\ldots\right). \quad (8.2.3)$$

Theorem 4.1.3 implies that $x(Z_n)$ given by (8.2.3) is a consistent with convergence in probability statistical decision procedure which is consistent if $L(\theta,x)$ is bounded for each θ.

The use of random elements in decision theory is not restricted to sequence spaces. Examples where the states Θ and alternatives X are subsets of other normed linear spaces can be handled similarly. Obviously, verifications of the conditions, restrictions on the problem, and the interpretations of the results depend on the particular space which is used. However, the above examples illustrate the usefulness of the weak laws of large numbers for random elements in defining consistent statistical decision procedures. Also, the strong laws of large numbers or convergence of weighted sums for random elements could have been used in this section but only convergence in probability was needed, and of course, the hypotheses for the weak laws are more easily satisfied.

8.3 APPLICATIONS IN QUALITY CONTROL

In Section 8.1 convergence results for random elements were used to estimate the drift function of Brownian motion processes in a manufacturing problem. In this section the use of weighted sums results in quality control problems is indicated.

Quality control is an important industrial application of statistics. In a continuous production process with parameter θ, samples of fixed size are taken at regular intervals of time. A statistic V_n is computed from the nth sample to detect the change in θ to determine when the process is out of control. The simplest and most widely used inspection scheme is the Shewhart (1931) control chart, where the V_n's are plotted on the chart and rectifying action is taken when the observed value of V_n falls outside the control limits drawn on the chart. In this single-sample scheme, none of the information of the previous samples is used. To achieve greater efficiency and make use of the previous observations, Dudding and Jennett (1944) proposed a control chart with warning lines within the control limits, where rectifying action is taken if any point falls outside the control limits or if s out of a sequence of t consecutive points fall outside the warning lines

A further extension may be made by considering the control charts based on the weighted sum

$$S_n = \sum_{k=1}^{n} a_{nk}|V_k|,$$

where corrective actions are taken when $S_n \geq d$, a preassigned constant. For example, the weights defined by $a_{nn} = 1$, and $a_{nk} = 0$ for $k < n$ generate the Shewhart chart. The control chart with warning lines can be obtained by considering $a_{nk} = d/s$ for $k = n, n-1, \ldots, n-t+1$ and $= 0$ otherwise, and

$$S_n = \max \{\sum_{k=1}^{n} a_{nk} I_{[|V_k| \geq h]}, |V_n|\},$$

where h is the distance from the warning line to the parameter θ (usually taken to be the center line).

Lai (1974) considered the case when $a_{nk} = c_{n-k}$, where $c_0 \geq c_1 \geq \ldots \geq c_{i-1} > 0 = c_i = c_{i+1} = \ldots$. This sequence puts heavy weights on the immediate past and little weight on the remote past. Generally, this weighted sum scheme is difficult to evaluate. Lai (1974) has considered the average run length (ARL), $E_\theta N$, as a measure of the goodness of the above scheme, where ARL is the expected number of articles sampled before action is taken when θ has remained at a constant level. The limiting behavior of the sequence $\{S_n\}$ (such as the convergence results of Chapter III or Chapter V) may be applied in the study of the asymptotic behavior of the average run length.

8.4 APPLICATIONS IN ESTIMATION PROBLEMS

In this section some applications in estimation problems are considered. First, a generalized Monte Carlo method of Halton (1970) is described in a Fréchet space. Next, the consistency of the M-statistics of Huber (1964) is related to the convergence of weighted sums. Finally, a function space approach to the density estimation problem is given. It seems very appropriate to end this chapter on the density estimation problem since it illustrates the basic theme in these applications. Specifically, the direct application of the random elements results may yield less efficient results than particular techniques developed for the specific problem. However, the adaption (and extension) of the function space techniques for specific problems yeilds very powerful tools.

For the first application, consider the following generalization of the Monte Carlo method by Halton (1970). Suppose θ is the solution of some problem (such as the value of an integral). The usual Monte Carlo procedure is to express the solution θ, a real number, as the expected value of a random variable Z with finite variance defined on a probability space (Ω,A,P). Then points $\omega_1,\omega_2,\ldots$ are sampled independently from Ω and an estimator $\phi_n = \frac{1}{n}\sum_{k=1}^{n} Z(\omega_k)$ is formed so that $E[\phi_n] = \theta$. By the laws of large numbers for random variables, the sequence $\{\phi_n\}$ converges to θ in some mode (in probability or with probability one). Now suppose that θ is a point in a Fréchet space F as defined in Chapter I (for example, θ may be the solution of a functional equation). A sequence of estimators $\{V_n\}$ of θ may be formed in a manner similar to that for random variables, where V_n is a random element in F for each n with $EV_n = \theta$. If $\{V_n\}$ converges to θ in some mode (such as in probability or with probability one), then $\{V_n\}$ is defined to be a Monte Carlo process for θ. Thus, the characterizations in Chapter II and the results of Chapters IV, V, and VI can be applied to obtain a convergence theory for this process.

Next, Huber (1964) has introduced a class of statistics (known as M-statistics) from a sample $\{y_k\}$ for the center of a symmetric distribution which are solutions T of the equation

$$\sum_{k=1}^{n}\Psi\left(\frac{y_k-T}{s}\right) = 0, \qquad (8.4.1)$$

where Ψ is an increasing odd function and s is a robust measure of the spread of the distribution. This concept arises from the techniques of finding maximum likelihood estimators, MLE's. The MLE's have been shown nearly optimal in sampling 10 - 20 observations from a known distribution (see Hoaglin (1971)). However, in most cases, such optimality is confined to the particular distribution, while the M-estimates retain the near-optimality of the MLE but are not restricted to a particular density.

One of the most intuitive computational approaches to solving (8.4.1) is based on forming a weighting function

$$\Psi(z) = w(z)\cdot z. \tag{8.4.2}$$

Writing $Y_k = (y_k - T)/s$, these lead to

$$T_n = \sum_{k=1}^{n} [w(Y_k)/\sum_{j=1}^{n} w(Y_j)]y_k, \tag{8.4.3}$$

a sequence of weighted sums of random variables. Thus, the consistency of the M-estimates is an immediate result of the convergence of such weighted sums.

In the density estimation problem, let $\{X_i\}$ be independent observations from a continuous but unknown probability density functions, f(t). An intuitive approach (even to a layman) would be the frequency histogram or the computation of

$$f_n(t) = \frac{1}{n} \sum_{i=1}^{n} \sum_{j=-\infty}^{\infty} I_{[j-1<X_i\le j]} I_{(j-1,j]}(t) \tag{8.4.4}$$

Since each $t \in (j_0-1, j_0]$ for some j_0,

$$f_n(t_0) = \frac{1}{n} \sum_{i=1}^{n} I_{[j_0-1<X_i\le j_0]} \to P[j_0 - 1 < X_1 \le j_0]$$

with probability one. Thus, for each $t \in R$,

$$f_n(t) \to \sum_{j=-\infty}^{\infty} P[j-1 < X_1 \le j] I_{(j-1,j]}(t) = g(t)$$

with probability one. Thus, the approximation where g(t) is the

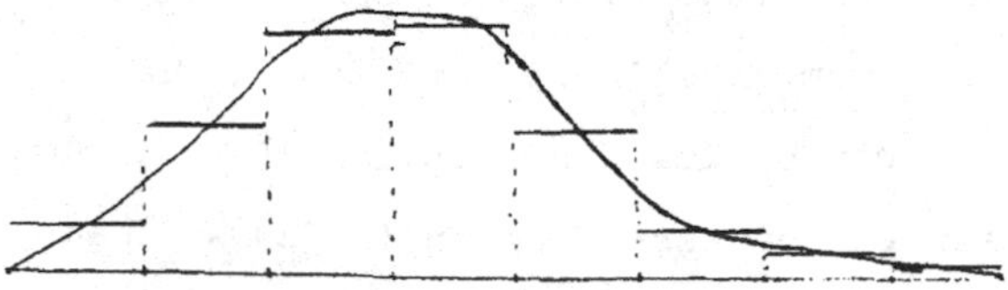

"step-like" function and f(t) is smooth curve. Moreover, the subintervals could be made smaller and a polygonal approximation could be used for a closer fit.

For a function space approach (suggested by Ramon Leon), let

$$C_0 = \{f: \text{ f is a continuous function from R into R and } \lim_{x\to\infty} f(x) = 0 = \lim_{x\to-\infty} f(x)\} \tag{8.4.5}$$

For $f \in C_0$, define $||f|| = \sup_t |f(t)|$. Thus, C_0 is a normed linear space which includes the uniformly continuous probability density functions. In (8.4.4),

$$\sum_{j=-\infty}^{\infty} I_{[j-1<X_i\le j]} I_{[j-1<t\le j]}(t)$$

placed the uniform probability density on (j-1,j] when $X_i \in (j-1,j]$. Thus, for each k let

$$T_k(t) = \sum_{j=-\infty}^{\infty} f_{k,j}(t) I_{(\frac{j-1}{2^k},\frac{j}{2^k}]}(t) \tag{8.4.6}$$

where $\{f_{k,j}(t)\}$ are continuous probability density functions with support $(\frac{2j-3}{2^{k+1}},\frac{2j+1}{2^{k+1}})$. Thus, T_k is a Borel measurable function from R into C_0 since it is countably-valued and $(\frac{j-1}{2^k},\frac{j}{2^k}] \in B(R)$. Also, $\{T_k(X_i): i = 1,2,3,\ldots\}$ are independent, identically distributed random elements in C_0, and $T_k(X_i)$ places the probability density function $f_{k,j}$ on the interval $(\frac{2j-3}{2^{k+1}},\frac{2j+1}{2^{k+1}})$ when $X_i \in (\frac{j-1}{2^k},\frac{j}{2^k}]$. Thus,

$$f_{nk}(t) = \frac{1}{n}\sum_{i=1}^{n} T_k(X_i) \to ET_k(X_1) = \sum_{j=-\infty}^{\infty} f_{k,j}(t) P[\frac{j-1}{2^k} < X_1 \le \frac{j}{2^k}]$$

with probability one by Theorem 4.1.1 under suitable moment conditions. Since $||f|| = \sup |f(t)|$,

$$\sup_t \left|\frac{1}{n}\sum_{i=1}^{n}\sum_{j=-\infty}^{\infty} f_{k,j}(t)I_{[\frac{j-1}{2^k}<X_i\leq\frac{j}{2^k}]}\right.$$

$$\left. - \sum_{j=-\infty}^{\infty} f_{k,j}(t)P[\frac{j-1}{2^k} < X_1 \leq \frac{j}{2^k}]\right|$$

goes to zero with probability one uniformly.

For a particular choice, let $f_{k,j}(t)$ be triangular on $(\frac{2j-3}{2^{k+1}}, \frac{2j+1}{2^{k+1}})$ with a peak of 2^k at $\frac{2j-1}{2^{k+1}}$. Then, for each k

$$||f_{n,k}(t) - ET_k(X_1)|| \to \infty$$

with probability one. In this case $ET_k(X_1)$ is polygonal on the points $\{\frac{2j-1}{2^{k+1}} : j = 0, \pm1, \pm2, \ldots\}$ and

$$\begin{aligned} ET_k(X_1)(\frac{2j-1}{2^{m+1}}) &= 2^k P[\frac{j-1}{2^k} < X_1 \leq \frac{j}{2^k}] \\ &= 2^k \int_{(j-1)/2^k}^{j/2^k} f_{X_1}(t)dt = f_{X_1}(t_0) \end{aligned} \tag{8.4.7}$$

for some $t_0 \in (\frac{j-1}{2^k}, \frac{j}{2^k}]$ by the mean-value theorem. Thus, a uniform approximation of arbitrary closeness [by the uniform continuity of f] is assured with probability one.

Researchers in the area of density estimation, use the estimate

$$f_n(t) = \frac{1}{nb(n)} \sum_{k=1}^{n} W(\frac{t-X_k}{b(n)})$$

where W is a weight function and $\{b(n)\}$ is a sequence of bandwidths converging to 0 [see Rosenblatt (1971) for a general discussion of curve estimate]. By adapting the function space techniques and using sub-Gaussian properties, Taylor and Cheng (1977) obtained the complete convergence of

$$\sup_{-\infty<t<\infty} \left|\frac{1}{nb(n)} \sum_{k=1}^{n} W(\frac{t-X_k}{b(n)}) - f(t)\right|$$

to zero under varying conditions on the bandwidths b(n) and smoothness of the weight function W.

BIBLIOGRAPHY

Ahmad, S. (1965). Eléments aléatoires dans les espaces vectoriels topologiques. Ann. Inst. Henri Poincaré Sect. B, 2, 95-135.

Alf, Carol (preprint). Rates of convergence of laws of large numbers for independent Banach-valued random variables.

Anderson, T.W. and Taylor, J.B. (1976). Strong consistency of least squares estimates in normal linear regression. Ann. Statist. 4, 788-790.

Baciu, A. (1971). La loi des grands nombres pour variables faiblement correlées. An. Univ. Bucuresti, Mat.-Mec. 20, 9-17.

Beck, A. (1963). On the strong law of large numbers. Ergodic Theory. Academic Press, New York, 21-53.

Beck, A. (1976). Cancellation in Banach spaces. Probability in Banach Spaces, Oberwolfach 1975, LECTURE NOTES IN MATHEMATICS, V526, Springer-Verlag, Berlin, 13-20.

Beck, A. and Giesy, D. P. (1970). P-uniform convergence and a vector-valued strong law of large numbers. Trans. Amer. Math. Soc. 147, 541-559.

Beck, Giesy, and Warren (1975). Recent developments in the theory of strong laws of large numbers for vector-valued random variables. Theory of Probability and Appl., XX, 127-134.

Beck, A. and Warren, P. (1968). A strong law of large numbers for weakly orthogonal sequences of Banach-space random variables. MRC Technical Summary Report #848, University of Wisconson.

Beck, A. and Warren, P. (1972). Weak orthogonality. Pacific J. Math. 41, 1-11.

Beck, A. and Warren, P. (1974). Strong laws of large numbers for weakly orthogonal sequences of Banach space-valued random variables. Annals of Probability, 2, 918-925.

Beck, A. and Warren, A. (to appear). Counterexamples to strong laws of large numbers for Banach space-valued random variables. Asterique.

Billingsley, P. (1968). Convergence of Probability Measures. Wiley, New York.

Billingsley, P. (1974). A note on separable stochastic processes. Annals of Probability, 2, 476-479.

Binmore, K. G. and Katz, M. (1968). A note on the strong law of large numbers. Bull. Amer. Math. Soc. 74, 941-943.

Blanc-Lapierre, A. and Tortrat, A. (1968). Sur la loi forte des grands nombres pour les fonctions aléatoires stationnaires du second ordre. C. R. Acad. Sci. Paris. Ser. A 267, 740-743.

Blanc-Lapierre, A. and Tortrat, A. (1970). Loi forte des grands nombres pour les fonctions aleatories stationnaires d'ordre deux et changement aléatorie d'horlage. C. R. Acad. Sci. Paris, Ser. A 270, 1186-1189.

Chatterji, S. D. (1964). A note on the convergence of Banach-space valued martingales. Math. Annalen, 153, 142-149.

Chatterji, S. D. (1976). Vector-valued martingales and their applications. Probability in Banach Spaces, Oberwolfach 1975, LECTURE NOTES IN MATHEMATICS, V526, Springer-Verlag, Berlin, 33-51.

Chow, Y. S. (1966). Some convergence theorems for independent random variables. Annals of Math. Statist, 35, 1482-1493.

Chow, Y. S. (1967). On a strong law of large numbers for martingales Ann. Math. Statist. 38, 610.

Chow, Y. S. and Lai, T. L. (1973). Limiting behavior of weighted sums of independent random variables. Ann. Prob., 1. 810-824.

Chung, K. L. (1947). Note on some laws of large numbers. Amer. J. Math., 69, 189-192.

Chung, K. L. (1968). A Course in Probability. Harcourt, Brace and World, New York.

Cohn, Harry (1970). On the strong law of large numbers for a class of dependent random variables. Revue Roumaine Math. Pur. Appl. 15, 487-493.

Csörgö, Miklós (1968). On the strong law of large numbers and the central limit theorem for martingales. Trans. Amer. Math. Soc. 131, 259-275.

Daffer, P. Z. and R. L. Taylor (1977). Laws of large numbers for D[0,1]. USC Mathematics Technical Report No. 60B05-10 University of South Carolina (to appear in the Annals of Probability).

DeAcosta, A. D. (1970). Existence and convergence of probability measures in Banach spaces. Trans. Amer. Math. Soc. 152, 273-298.

Delporte, Jean (1963). Un critère de convergence forte presque sûre des sommes d'éléments aléatoires indépendents dans un espace de Banach. Calcul Des Probabilites C. R. Acad. Sc. 257, 35-37.

Delporte, J. (1964). Fonctions aléatoires presque sûrement continues sur un intervalle ferme. Ann. Inst. Henri Poincaré, Vol I. 111-215.

Deo, C. M. and Truax, D. R. (1968). A note on the weak law. Ann. Math. Statist. 39, 2159-2160.

Doob, J. L. (1953). Probability in function space. Bull. Amer. Math. Soc. 53, 15-30.

Doob, J. L. (1947). Stochastic Processes. Wiley, New York.

Driml, M. (1960). Convergence of compact measures on metric spaces. Trans. Second Prague Conf. on Information Theory, Statist. Decision Functions, and Random Processes (1959), 71-92.

Driml, M. and Hans, O. (1960). Conditional expectations for generalized random variables. Trans. Second Prague Conf. on Information Theory, Statist. Decision Functions, and Random Processes (1959), 123-143.

Dudding, B. P. and Jennett, W. J. (1944). Quality Control Chart Technique. General Electric, London.

Dudley, R. M. (1970). Random linear functionals: Some recent results. Lectures in Mod. Anal. & Appl., III, 62-70. Lecture Notes in Math., V170. Springer, Berlin.

Dudley, R. M. (to appear). Speeds of metric convergence. Z. Wahr. Verw. Gebiete.

Dudley, Feldman, & LeCam (1971). On seminorms and probabilities, and abstract Wiener spaces. Ann. of Math. 93, 390-408.

Egorov, V. A. (1970). Some theorems on the strong law of large numbers and the law of the iterated logarithm. Doklady Akad. Nauk. SSSR 193, 268-271.

Egorov, V. A. (1970). On the strong law of large numbers and the law of the iterated logarithm for sequences of independent random variables. Theory Probab. Appl. 15, 509-514.

Erdös, P. and Rényi, A. (1970). On a new law of large numbers. J. Analyse Math. 23, 103-111.

Fernandez, P. J. (1971). A weak convergence theorem for random sums in a normed space. Annals Math. Statist. 42, 1737-1741.

Fortet, R. and Mourier, E. (1955). Les fonctions aléatoires comme elements aléatoires dan des espaces de Banach. Studia Math 55, 62-79.

Franck, W. E. and Hanson, D. L. (1966). Some results giving rates of convergence in the law of large numbers for weighted sums of independent random variables. Trans. Amer. Math. Soc., 124, 347-359.

Garsia, A., Posner, E., and Rodemich, E., (1968). Some properties of the measures on function spaces induced by Gaussian processes. J. Math. Anal. and Appl. 21, 150-161.

Giesy, D. P. (1965). On a convexity condition in normed linear spaces. Trans. Amer. Math. Soc. 125, 114-146.

Giesy, D. (1976). Strong laws of large numbers for independent sequences of Banach space-valued random variables. Probability in Banach spaces, Oberwolfach 1975, LECTURE NOTES IN MATHEMATICS, V526, Springer-Verlag, Berlin, 89-99.

Glivenko, V. (1928). Sur la loi des grands nombres dans les espaces fonctionnels. Rend. Lincei 8, 673-676.

Govindarajulu, Z. (1970). On weak laws of large numbers. Proc. Indian Acad. Sci., Sect. A 71, 266-274.

Grenander, U. (1963). Probabilities on Algebraic Structures. Wiley, New York.

Halton, J. H. (1970). A retrospective and prospective survey of the Monte Carlo method. SIAM Review 12, 1-63.

Hahn, M. (1975). Central Limit Theorems for D[0,1]-valued Random Variables. Ph.D. Dissertation, MIT.

Hanš, O. (1957). Generalized random variables. Trans. First Prague Conf. on Information Theory, Statist. Decision Functions, and Random Processes (1956), 61-103.

Hanson, D. L. and Pledger, G. (1969). On the mean ergodic theorem for weighted averages. Z. Wahr verw. Geb, 13, 141-149.

Hanson, D. L. and Wright, F. T. (1971). Some convergence results for weighted sums of independent random variables. Z. Wahr verw. Geb., 19, 81-89.

Halmos, P. R. (1950). Measure Theory. Van Nostrand, Princeton.

Heyde, C. C. and Rohatgi, V. K. (1967). A pair of complementary theorems on convergence rates in the law of large numbers. Proc. Conf. Soc., 63, 73-82.

Hille, E. and Phillips, R. (1957). Functional Analysis and Semi-groups. Amer. Math. Soc. Colloq. Publ. Vol. XXXI. Providence, Rhode Island.

Hoaglin, D. (1971). Optimal invariant estimation of location for three distributions and the invariant efficiency of some other estimations. Ph.D. Thesis, Princeton University.

Hoffmann-Jørgensen, J. (1974). Sums of independent Banach space valued random variables. Studia Math., J. LII., 159-188.

Hoffmann-Jørgensen, J. and Pisier, G. (1976). The law of large numbers and the central limit theorem in Banach spaces. Annals of Probability, 4, 587-599.

Horváth, J. (1966). Topological Vector Spaces and Distributions. Vol. 1, Addison-Wesley, Reading, Mass.

Huber, P. J. (1964). Robust estimation of a location parameter. Ann. Math. Statist., 35, 73-101.

Itô, K., and Nisio, M. (1968). On the convergence of sums of independent Banach space valued random variables. Osaka J. Math. 5, 35-48.

Itô, K. (1970). Canonical measurable random functions. Proc. Internat. Conf. on Functional Anal. & Related Topics, 369-377. Univ. of Tokyo Press, Tokyo.

Itô, K. and McKean, H. P. (1974). Diffusion Processes and their Sample Paths. Springer-Verlag, Berlin, Heidelberg, and New York

Jain, N. C. (1976). An example concerning CLT and LIL in Banach spaces. Annals of Probability, 4, 690-694.

Jajte, R. (1968). On stable distributions in Hilbert spaces. Studia Math. 30, 63-71.

Jajte, R. (1968). On the probability measures in Hilbert spaces. Studia Math 29, 221-241.

Jajte, R. (1968). On convergence of infinitely divisible distributions in a Hilbert space. Colloq. Math. 19, 327-332.

James, R. C. (1974). A non-reflexive Banach space that is uniformly nonoctrahedral. Israel J. Math., 18, 145-155.

Jamison, Orey, and Pruitt (1965). Convergence of weighted averages of independent random variables. Z. Wahr. verw. Geb., 4, 40-44.

Jouandet, O. (1970). Sur la convergence en type de variables aléatoires à valeurs dans des espace d'Hilbert ou de Banach. C. R. Acad. Sci. Paris Ser. AB271, A1082-A1085.

Kannan, D. and Bharucha-Reid, A. T. (1970). Note on covariance operators of probability measures on a Hilbert space. Proc. Japan Acad. 46, 124-129.

Kannan, D. (1973). Martingales in Banach spaces with Schauder bases. J. Math. & Phy. Sciences, VII, 93-100.

Katz, Melvin (1968). A note on the weak law of large numbers. Ann. Math. Statist. 39, 1348-1349.

Kelley and Namioka (1963). Linear Topological Spaces. Van Nostrand, Princeton.

Kruglov, V. M. (1971). Convergence of distributions of sums of independent random variables with values in Hilbert space to a normal and a Poisson distributions. Soviet Math. Dokl., 12, 661-664.

Kruglov, V. M. (1971). Convergence of the distributions of sums of independent random variables with values in Hilbert space. Theory Prob. Appl. 16, 350-351.

Kuelbs, J. (1970). Gaussian measures on a Banach space. J. Funct. Anal. 5, 354-367.

Kuelbs, J. and Mandrekar, V. (1970). Harmonic analysis in certain vector spaces. Trans. Amer. Math. Soc. 149, 213-231.

Kuelbs, J. and Mandrekar, V. (1972). Harmonic analysis on F-spaces with a basis. Trans. Amer. Math. Soc., 169, 113-152.

Kuelbs, J. (1976). A strong convergence theorem for Banach space valued random variables. Annals of Probability 4, 744-771.

Kuelbs, J. (1976). A counterexample for Banach space valued random variables. Annals of Probability 4, 684-689.

Lai, T. L. (1972). Control charts based on weighted sums. Ann. Statist. 2, 134-147.

Lai, T. L. (1974). Convergence rates in the strong law of large numbers for random variables taking values in Banach spaces. Bull. Inst. Math. Acad. Sinica, 2, 67-85.

LeCam, L. (1957). Convergence in distribution of stochastic processes. Univ. California Publ. Statist, 2, 207-236.

Liggett, T. M. (1970). Weak convergence of conditional sums of independent random vectors. Trans. Amer. Math. Soc. 152, 195-213.

Lingren, B. W. (1963). Statistical Theory, MacMillan, New York.

Loéve, M. (1963). Probability Theory. Van Nostrand, Princeton.

Loynes, R. M. (1969). The consistency of certain sequential estimators. Annals of Math. Statist., 40, 568-574.

Mandekar, V. and Mann, H. B. (1969). On the realization of stochastic processes by probability distributions in function spaces. Sankhyā, Ser. A 31, 477-480.

Mangano, G. C. (1976). Sequential compactness of certain sequences of Gaussian random variables with values in C[0,1]. Ann. Prob., 4, 902-913.

Mann, H. B. (1953). On the realization of stochastic processes by probability distribution in function spaces. Sankhyā 11, 3-8.

Marti, J. T. (1969). Introduction to the Theory of Bases. Springer-Verlag, New York.

Mourier, E. (1953). Eléments aleatories dan un espace de Banach. Ann. Inst. Henri Poincaré 13, 159-244.

Mourier, E. (1956). L-random elements and L*-random elements in Banach spaces. Proc. Third Berkeley Sympos. Math. Statist. and Prob. 2, 231-242.

Mourier, E. (1967). Random elements in linear spaces. Proc. 5th Berkeley Sympos. Math. Statist. Prob., Univ. of Cal. 1965/1966 2, Part 1, 43-53.

Muštari, D. H. (1971). On almost sure convergence in linear spaces of random variables. Theory Probab. Appl. 15, 337-342.

Nagaev, S. V. (1972). On necessary and sufficient conditions for the strong law of large numbers. Theory Prob. & Appl., 17, 573-581.

Nedoma, J. (1957). Note on generalized random variables. Trans. First Prague Conf. Inform. Theory, Statistical Decision Functions, and Random Processes, Prague, 139-141.

Negishi, H. (1975). On the convergence of sums of independent random elements in some Fréchet space. Science Reports, Yokohama Univ. #22, 11-18.

Neveu, J. (1965). Mathematical Foundations of the Calculus of Probability. Holden-Day, San Francisco.

Padgett, W. J. and Taylor, R. L. (1973). Laws of Large Numbers for Normed Linear Spaces and Certain Fréchet Spaces. LECTURE NOTES IN MATHEMATICS, Vol. 360, Springer-Verlag, Berlin.

Padgett, W. J. and Taylor, R. L. (1974). Convergence of weighted sums of random elements in Banach space and Fréchet spaces. Bull. Inst. Math., Acad. Sinica 2, 389-400.

Padgett, W. J. and Taylor, R. L. (1976). Almost sure convergence of weighted sums of random elements in Banach spaces. Probability in Banach Spaces, Oberwolfach 1975, LECTURE NOTES IN MATHEMATICS, V526, Springer-Verlag, Berlin, 187-202.

Parthasarthy, K. R. (1967). Probability Measures on Metric Spaces. Academic Press, New York and London.

Perlman, M. (1972). Characterizing measurability, distribution, and weak convergence of random variables in a Banach space by total subsets of linear functionals. J. of Mult. Anal., 2, 174-188.

Petrov, V. V. (1969). On the strong law of large numbers. Teor. Verojatn. Primen. 14, 193-202.

Pettis, B. J. (1938). On integration in vector spaces. Trans. Amer. Math. Soc. 44, 277-304.

Pisier, G. (1974). Sur les espaces qui ne contiennent pas de ln^1 uniformément. Sem. Maurey-Schwartz, Ecole Polytechnique, Paris, p. VII. 1 - VII. 19.

Pop-Stojanovic, Z. R. (1971). On the strong law of large numbers for Banach-valued weakly integrable random variables. J. Math. Soc. Japan 23, 269-277.

Prohorov, Yu. V. (1956). Convergence of random processes and limit theorems in probability theory. Theory Prob. Appl. 1, 157-214.

Prohorov, Yu. V. (1956). The method of characteristic functionals. Proc. Fourth Berk. Symp. on Math Stat. & Prob. 2, 403-418.

Prohorov, Yu. V. (1959). Some remarks on the strong law of large numbers. Theory Prob. & Appl., 4, 204-208.

Pruitt, W. (1966). Summability of independent random variables. J. Math. & Mech. 15, 769-776.

Révész, P. (1968). The Laws of Large Numbers. Academic Press, New York.

Riečan, B. (1970). A general form of the law of large numbers. Acta Fac. Rer. Natur. Univ. Comenian., Math. 24, 129-138.

Rao, R. Ranga (1963). The law of large numbers for D[0,1]-valued random variables. Theory of Prob. & Appl. 8, 70-74.

Rohatgi, V. K. (1968). Convergence rates in the law of large numbers II. Proc. Cambridge Philos, Soc. 64, 485-488.

Rosenblatt, M. (1971). Curve estimates. Annals Math. Statist., 42, 1815-1842.

Rosenkrantz, W. (1969). A rate of convergence for the Von Mises statistic. Trans. Amer. Math. Soc., 139, 329-337.

Royden, H. L. (1972). Real Analysis, 2nd Ed., MacMillan, New York.

Rudin, W. (1973). Functional Analysis. McGraw-Hill, New York.

Scalora, F. S. (1961). Abstract martingale convergence theorems. Pacific J. Math., 11, 347-374.

Schaefer, H. H. (1966). Topological Vector Spaces. MacMillan, New York.

Shewhart, W. (1931). Economic Control of Quality of Manufactured Product. Van Nostrand, Princeton.

Slivka, John and Severo, N. C. (1970). On the strong law of large numbers. Proc. Amer. Math. Soc. 24, 729-734.

Skorokhod, A. V. (1970). Gaussian measures in Banach spaces. Theory of Prob. & Appl. 15, 588.

Smythe, R. T. (1973). Strong laws of large numbers for r-dimensional arrays of random variables. Annals of Probability 1, 164-170.

Stojanović, Steven M. (1966). On the law of large numbers for generalized stochastic processes. Mat. Vesnik., n. Ser. 3 (18) 299-302.

Stout, W. (1968). Some results on the complete and almost sure convergence of linear combinations of independent random variables and martingale differences. Annals of Math Statist., 39, 1549-1562.

Stout, W. F. (1974). Almost Sure Convergence. Academic Press, New York.

Steiger, W. L. (1973). Weak laws for dependent sums. Proc. Amer. Math. Soc. 41, 278-281.

Szynal, Dominik (1962). On the strong law of large numbers for random variables bounded by sequences of numbers. Ann. Univ. Mariae Curie-Sklodowska, Sect. A 16, 123-127.

Szynal, Dominik (1964). A note on qualitative conditions for the strong law of large numbers. Ann. Univ. Mariae Curie-Sklodowska, Sect. A 18, 5-7 (1968).

Taylor, A. E. (1958). Introduction to Functional Analysis. Wiley, New York.

Taylor, R. L. (1972). Weak laws of large numbers in normed linear spaces. Ann. Math. Statist. 43, 1267-1274.

Taylor, R. L. (1973). Consistent multi-attributed decision procedures. Multiple Criteria Decision Making. Univ. of S.C., October, 1973, 774-777.

Taylor, R. L. (1974). Convergence of elements in random normed spaces. Bull. Austral. Math. Soc. 12, 31-47.

Taylor, R. L. and Padgett, W. J. (1974). Some laws of large numbers for normed linear spaces. Sankhya, 37, Series A, 359-368.

Taylor, R. L. and Padgett, W. J. (1975). Stochastic convergence of weighted sums in normed linear spaces. J. Multivariate Analysis, Vol. 5, No. 4, 434-450.

Taylor, R. L. and Padgett, W. J. (1976). Weak laws of large numbers in Banach spaces and their extensions. Probability in Banach Spaces, Oberwolfach 1975, LECTURE NOTES IN MATHEMATICS, V526, Springer-Verlag, Berlin, 227-242.

Taylor, R. L. (1976). Properties of stochastic processes characterized as function-valued random variables. USC Mathematics Technical Report No. 60B05-7, University of South Carolina.

Taylor, R. L. and Daffer, P. (to appear). Some weak laws of large numbers for probability measures on vector spaces. Proc. Conf. on Vector Space Measures. LECTURE NOTES IN MATHEMATICS.

Taylor, R. L. and Cheng, K. F. (1977). On the uniform complete convergence of density function estimates. FSU Statistics Report M446. Florida State University, December 1977.

Taylor, R. L. and Wei, D. (to appear). Laws of large numbers for tight random elements in normed linear spaces. Annals of Probability.

Tucker, H. G. (1967). A Graduate Course in Probability. Academic Press, New York.

Varadarajan, W. S. (1961). Convergence of stochastic processes. Bull. Amer. Math. Soc. 77, 276-280.

Veršik, A. M. (1966). Duality in the theory of measures in linear spaces. Soviet Math. 7, 1210-1213.

Wagner, T. J. (1969). On the rate of convergence for the law of large numbers. Ann. Math. Statist. 40, 2105-2107.

Walsh, J. B. (1967). A note on uniform convergence of stochastic processes. Proc. Amer. Math. Soc. 18, 129-132.

Warren, P. and Howell, J. (1976). A strong law of large numbers for orthogonal Banach space-valued random variables. Probability in Banach spaces, Oberwolfach 1975, LECTURE NOTES IN MATHEMATICS, V526, Springer-Verlag, Berlin, 253-262.

Wei, D. and Taylor, R. L. (to appear, 1978a). Geometric consideration of weighted sums convergence and random weighting. Bull. Inst. Math. Acad. Sinica.

Wei, D. and Taylor, R. L. (to appear, 1978b). Convergence of weighted sums of tight random elements. Journal of Multivariate Analysis.

Whitt, W. (1970). Weak convergence of probability measures on the function space C[0,∞). Ann. Math. Statist. 41, 939-944.

Wichura, M. J. (1971). A note on the weak convergence of stochastic processes. Annals Math. Statist. 42, 1769-1772.

Wilansky, A. (1964). Functional Analysis. Blaisdell, New York.

Willard, S. (1968). General Topology, Addison-Wesley, Reading, Mass.

Woyczynski, W. A. (1973). Random series and law of large numbers in some Banach spaces. Theory of Probability and Appl., 18, 350-355.

Woyczynski, W. A. (1976). Asymptotic behavior of martingales in Banach spaces. Probability in Banach Spaces, Oberwolfach 1975, LECTURE NOTES IN MATHEMATICS, V526, Springer-Verlag Berlin, 273-284.

Wright, F. T., Platt, R. D., and Robertson, T. (1977). A strong law for weighted averages of independent, identically distributed random variables with arbitrarily heavy tails. Annals of Prob., 5, 586-590.

Yosida, K. (1965). Functional Analysis. Springer-Verlag, Berlin-Göttingen-Heidelberg.

SUBJECT INDEX

Banach space
- condition G_α 97,
- convex of type (B) 86,
- definition 9,
- the space c 10,
- the space c_0 10,
- the space C[0,1] 10,
- the space L^p 11,
- the space L^∞ 11,
- the space ℓ^p 10,
- type p 102,

Beck's Strong Law of Large Numbers for B-convex spaces 86-96,

Beck and Giesy's Strong Law of Large Numbers 82-83,

Bochner integral 39,

Borel
- function 12,
- subsets 12,
- subsets in D[0,1] 161,

Completeness 7,

Consistent decision procedure 191,
- consistent with convergence in probability 191,

Convergence
- almost surely 28,
- comparisons 29,
- complete 67,
- in D[0,1] 154-165,
- in mean 28,
- in probability 28,
- of series 105,
- of series of random elements in G_α spaces 99-100,
- of stochastic process 186-190,
- of weighted sums of random elements 108-145,
- of weighted sums of random variables 63-70,
- rates of 105,
- with probability one 28,

Convexity (also see Geometric Conditions)
- convex of type (B) 86,
- convexity and tightness in D[0,1] 166,179-184,
- discussion of 86-87,
- lack of convexity in D[0,1] 158,
- uniformly convex 86,

Determining class (see Family of unicity) 31,

D[0,1]
- basic properties 154-159,
- characterization of compactness 158,
- Daffer and Taylor's SLLN's 175-179,
- expected values 159,
- general introduction 4,
- Ranga Rao's SLLN 172-174,

topology of the Skorohod metric 154-155,
uniform topology 154-155,
Dual space
in characterizing identical distributions and independence 32-33,
in the convergence of weighted sums 110-111,
in the laws of large numbers 75-80,
of a linear topological space 8,
Family of unicity
definition 31,
examples 31-32,
Fréchet space
convergence of weighted sums 136,
definition 9,
laws of large numbers 104-106
the space $C[0,\infty)$ 11,
the space F 11,
the space s 9,
Generalized Gaussian random variables 66,
Geometric conditions
convergence of weighted sums 137-145,
convex of type (B) 86,
G_α condition 97,
type p 102-103,
the strong law of large numbers 96-104,
Hahn-Banach Theorem 13,
use of 27, 94,
Hilbert space
coordinate uncorrelation 61,
definition 9,
laws of large numbers 49-54
the space L^2 11,
the space ℓ^2 10,
Hoffmann-Jørgensen and Pisier's Strong Law of Large Numbers 102-103,
Identical Distributions
and transformations 30-31,
definition 30,
characterized by the dual space 32,
characterized by finite-dimensional distributions 34,
Independence
characterized by the dual space 32-33,
characterized by finite-dimensional distributions 36,
definition 30,
moment conditions in G_α spaces 97-98,
Inner product
definition 9,
in defining uncorrelation 50,

Ito and Nisio's convergence
of series 105,
Kronecker's lemma 101,
Isomorphism and Isomorphic 13,
Isometry and Isometric 13,
Laws of Large Numbers
and Geometric conditions 96-104,
Beck's SLLN and convexity 86-96,
Beck and Giesy's SLLNs 82-83,
Beck and Warren's SLLN 80,
counterexample in ℓ^1 80-81,
counterexample in ℓ^p 95,
convergence of weighted sums 3,
estimation applications 197-201,
general discussion 3,
Hoffman-Jørgensen and Pisier's
SLLN 102-103,
in a separable Hilbert
space 51-54,
in decision theory appli-
cations 190-193,
in Fréchet spaces 104-106,
in separable normed linear
spaces 71-104,
in type p spaces 102-103,
Monte Carlo methods 196-197,
Mourier's SLLN 72-74,
SLLN for tight random
elements 133,
SLLNs in D[0,1] 172-179,
strong laws for random
variables 46-48,
Taylor's WLLNs 75-79, 85,
Taylor and Padgett's SLLN 83-84,
weak law for random variables 45,
WLLN for tight random
elements 129,
WLLNs in D[0,1] 166-170,
Woycznski's SLLN 101-102,
Linear Space
definition 6,
the linear space D[0,1] 157,
Linear topological space 8,
Markov's inequality 29,
Measurability
Borel 12,
problem 1,
strongly 23,
Metric space
definition 7,
D[0,1] 157,
linear 8,
Mourier's Strong Law of
Large Numbers 72-74,
Norm
definition 8,
normed linear space 8,
of a continuous linear
functional 13,
uniform norm for D[0,1] 154.

Orthogonal
definition 18,
orthonormal 18,
weakly orthogonal 80,
Pettis integral
as an expected value 38,
existence of 40-41,
properties of 38-39,
Random element(s)
addition of 28,
characterized by the dual space 27-28,
characterized by a Schauder basis 24,
convergence of weighted sums 108-152,
countably-valued 22-23,
definition 21,
expected value of (see Pettis integral) 38,
in a separable Hilbert space 49,
in a separable semimetric space 25-26,
in D[0,1] 159-165,
in G_α spaces 97-104,
in R 21,
in R^n 21,
identically distributed 30,
independent 30,
laws of large numbers for 51-54, 71-104,
limits of 22-23,
probabilistic convergence 28,
product with a random variable 24,
strongly-measurable 23,
transformations of 22,
type (A) 88,
variance of 38,
uncorrelated 50,
Schauder basis
characterizing compactness 19,
characterizing random elements 24,
characterizing the expected value 42,
characterizing weakly uncorrelated 56,
coordinate functionals 14,
coordinate uncorrelation 55,
definition 14,
examples 15-18,
in convergence of weighted sums 112-127,
in obtaining laws of large numbers 72-75,
monotone 14,
partial sum operators 15,
Semimetric space 7,
Seminorm 8,
Separability
and random elements 23,

definition 9,
stochastic processes 2,
Skorohod metric 154,
Stochastic Processes
and random elements 1,
Brownian motion processes 36,
laws of large numbers
for 84, 186-190,
Poisson processes 36,
stationary increments and
identically distributed
random elements 35,
Taylor's Weak Laws of Large
Numbers 75-79,
Taylor and Padgett's Strong
Law of Large Numbers 83-84,
Tightness
centering random elements 121,
convergence of weighted sums
of random elements 124-134,
convex tightness in
D[0,1] 166, 179-184,
definition 120,
for Brownian motion processes 187,
Toeplitz sequence 63, 109,
Uncorrelation
and independence 52,
comparisons 55-63,
coordinate uncorrelated 55,
in a separable Hilbert space 50,
pointwise in D[0,1] 169,
random variables 45,
weakly uncorrelated 55,
Weighted Sums
applications in quality
control 195-196,
convergence and tightness 120-136,
convergence for random
elements 108-152,
convergence for random
variables 63-73,
convergence of randomly
weighted sums 146-151,
general appearance 2,
geometric conditions and
convergence 137-145,
random weighting 4,
Woycznski's Strong Law of Large
Numbers 101-102,